M000249967

The Palgrave Macmillan Animal Ethics Series

Series editors: **Andrew Linzey** and **Priscilla N. Cohn**

In recent years there has been a growing interest in the ethics of our treatment of animals. Philosophers have led the way and now a range of other scholars have followed, from historians to social scientists. From being a marginal issue, animals have become an emerging issue in ethics and in multidisciplinary inquiry. This series explores the challenges that animal ethics poses, both conceptually and practically, to traditional understandings of human–animal relations.

Specifically, the series will:

- provide a range of key introductory and advanced texts that map out ethical positions on animals;
- publish pioneering work written by new, as well as accomplished, scholars;
- produce texts from a variety of disciplines that are multidisciplinary in character or have multidisciplinary relevance.

Titles include:

ANIMALS AND PUBLIC HEALTH
Why Treating Animals Better Is Critical to Human Welfare
Aysha Akhtar

AN INTRODUCTION TO ANIMALS AND POLITICAL THEORY
Alasdair Cochrane

THE COSTS AND BENEFITS OF ANIMAL EXPERIMENTS
Andrew Knight

POPULAR MEDIA AND ANIMAL ETHICS
Claire Molloy

ANIMALS, EQUALITY AND DEMOCRACY
Siobhan O'Sullivan

SOCIAL WORK AND ANIMALS
A Moral Introduction
Thomas Ryan

AN INTRODUCTION TO ANIMALS AND THE LAW
Joan Schaffner

Forthcoming titles:

HUMAN ANIMAL RELATIONS
The Obligation to Care
Mark Bernstein

ANIMAL ABUSE AND HUMAN AGGRESSION
Eleonora Gullone

ANIMALS IN THE CLASSICAL WORLD
Ethical Perceptions
Alastair Harden

POWER, KNOWLEDGE, ANIMALS
Lisa Johnson

AN INTRODUCTION TO ANIMALS AND SOCIOLOGY
Kay Peggs

The Palgrave Macmillan Animal Ethics Series
Series Standing Order ISBN 978–0–230–57686–5 Hardback
978–0–230–57687–2 Paperback
(*outside North America only*)

You can receive future titles in this series as they are published by placing a standing order. Please contact your bookseller or, in case of difficulty, write to us at the address below with your name and address, the title of the series and the ISBN quoted above.

Customer Services Department, Macmillan Distribution Ltd, Houndmills, Basingstoke, Hampshire RG21 6XS, England

Animals and Public Health

Why Treating Animals Better Is Critical to Human Welfare

Aysha Akhtar

Oxford Centre for Animal Ethics, UK, and the US Food and Drug Administration, North Potomac, Maryland, USA

Libraries @ Becker College

palgrave
macmillan

© Aysha Akhtar 2012

All rights reserved. No reproduction, copy or transmission of this publication may be made without written permission.

No portion of this publication may be reproduced, copied or transmitted save with written permission or in accordance with the provisions of the Copyright, Designs and Patents Act 1988, or under the terms of any licence permitting limited copying issued by the Copyright Licensing Agency, Saffron House, 6–10 Kirby Street, London EC1N 8TS.

Any person who does any unauthorized act in relation to this publication may be liable to criminal prosecution and civil claims for damages.

The author has asserted her right to be identified as the author of this work in accordance with the Copyright, Designs and Patents Act 1988.

First published 2012 by
PALGRAVE MACMILLAN

Palgrave Macmillan in the UK is an imprint of Macmillan Publishers Limited, registered in England, company number 785998, of Houndmills, Basingstoke, Hampshire RG21 6XS.

Palgrave Macmillan in the US is a division of St Martin's Press LLC, 175 Fifth Avenue, New York, NY 10010.

Palgrave Macmillan is the global academic imprint of the above companies and has companies and representatives throughout the world.

Palgrave® and Macmillan® are registered trademarks in the United States, the United Kingdom, Europe and other countries.

ISBN 978–0–230–24973–8

This book is printed on paper suitable for recycling and made from fully managed and sustained forest sources. Logging, pulping and manufacturing processes are expected to conform to the environmental regulations of the country of origin.

A catalogue record for this book is available from the British Library.

A catalog record for this book is available from the Library of Congress.

10 9 8 7 6 5 4 3 2 1
21 20 19 18 17 16 15 14 13 12

Transferred to Digital Printing in 2012

To Dad (Abu), for teaching me the value of an education

Contents

Series Preface

This is a new book series for a new field of inquiry: animal ethics.

In recent years, there has been growing interest in the ethics of our treatment of animals. Philosophers have led the way and now a range of other scholars have followed, from historians to social scientists. From being a marginal issue, animals have become an emerging issue in ethics and in multidisciplinary inquiry.

In addition, a rethink of the status of animals has been fueled by a range of scientific investigations which have revealed the complexity of animal sentiency, cognition and awareness. The ethical implications of this new knowledge are yet to be properly evaluated, but it is becoming clear that the old view that animals are mere things, tools, machines or commodities cannot be sustained ethically.

But it is not only philosophy and science that are putting animals on the agenda. Increasingly, in Europe and the USA, animals are becoming a political issue as political parties vie for the 'green' and 'animal' vote. In turn, political scientists are beginning to look again at the history of political thought in relation to animals, and historians are beginning to revisit the political history of animal protection.

As animals grow as an issue of importance, so there have been more collaborative academic ventures leading to conference volumes, special journal issues and, indeed, new academic animal journals. Moreover, we have witnessed the growth of academic courses, as well as university posts, in animal ethics, animal welfare, animal rights, animal law, animals and philosophy, human–animal studies, critical animal studies, animals and society, animals in literature, and animals and religion—tangible signs that a new academic discipline is emerging.

'Animal ethics' is the new term for the academic exploration of the moral status of the non-human—an exploration that explicitly involves a focus on what we owe animals morally and that also helps us to understand the influences (social, legal, cultural, religious and political) that legitimate animal abuse. This series explores the challenges that animal ethics poses, both conceptually and practically, to traditional understandings of human–animal relations.

The series is needed for three reasons:

1. to provide the texts that will service the new university courses on animals;
2. to support the increasing number of students studying and academics researching in animal-related fields;
3. because there is currently no book series that is a focus for multidisciplinary research in the field.

Specifically, the series will

1. provide a range of key introductory and advanced texts that map out ethical positions on animals;
2. publish pioneering work written by new, as well as by accomplished, scholars;
3. produce texts from a variety of disciplines that are multidisciplinary in character or have multidisciplinary relevance.

The new *Palgrave Macmillan Animal Ethics* series is the result of a unique partnership between Palgrave Macmillan and the Ferrater Mora Oxford Centre for Animal Ethics, UK. It is an integral part of the mission of the center to put animals on the intellectual agenda by facilitating academic research and publication. The series is also a natural complement to one of the center's other major projects, the *Journal of Animal Ethics*. The center is an independent think tank for the advancement of progressive thought about animals, and it is the first of its kind in the world. It aims to demonstrate rigorous intellectual inquiry and the highest standards of scholarship. It strives to be a world-class center of academic excellence in its field.

We invite academics to visit the center's website at www. oxfordanimalethics.com and to contact us with new book proposals for the series.

Andrew Linzey and *Priscilla N. Cohn*
Series Editors

Acknowledgments

A good friend of mine who is an author once told me that writing is a lonely occupation. It certainly can be, but I was fortunate enough to be surrounded by incredibly smart family members, friends and colleagues who continually served as sounding boards and allowed me to step outside of my own thoughts throughout the writing of this book. I would never have been able to formulate the ideas presented here without my friend Greg Goodale. As a rhetorician, Greg challenged me to fine-tune my arguments and constantly helped me get 'unstuck' whenever I had a mental block (which was not infrequent). My sister Jabeen, an accomplished author, provided useful suggestions to 'jazz up' the writing style. Allison George had the miserable task of reviewing, editing and cleaning up all of my draft chapters. Her help is truly appreciated. My sister Sahar and brother-in-law William Wojtach, both philosophers, helped make sure that my philosophical arguments were sound and my coverage of the topic accurate. Tom Beauchamp also provided extremely helpful comments on this topic.

Dr John Pippin's brilliant insights into medical research have greatly influenced my chapter on animal experimentation. His review of that chapter was immensely helpful. Dr Andrew Knight's review of the animal experimentation chapter (Chapter 6) is also greatly appreciated. Justin Goodman, Thomas Hartung, Alka Chadna, Ray Greek, Andre Menache, Jarrod Bailey and Jonathan Balcombe were all wonderful in providing answers or sharing their research when I desperately needed facts and figures. Andrew Linzey, Priscilla Cohn, and my editors and liaisons at Palgrave Macmillan – Priyanka Gibbons, Melanie Blair and Cherline Daniel – allowed me the tremendous opportunity of publishing this book and guided me throughout the process. I must especially thank Andrew Linzey, who bamboozled me into writing the book in the first place. Without his urging, I would never have considered it. Even though I cursed his name at times (lovingly, of course) when I was sick of my own writing, I can now say that he was right. This book needed to be written.

While working on this book, I had very little time for my husband Patrick. He was always gracious, though, in allowing me the space and time I needed to meet my deadline. He even took over my share of the household chores to help free up my time, including making the bed (which is a big deal for him!). Patrick's support and our daily conversations about my book helped keep me focus on the main theme. Most of all, I am so very, very grateful for the daily humor he has brought into my life.

I am also thankful to my friends and colleagues in public health who work tirelessly and often thanklessly to alleviate human suffering. I hope that this book helps them to also recognize the tremendous suffering we cause to animals and how their poor treatment in turn harms human health. Perhaps then we can start to find improved ways to better the lives of humans and non-human animals alike.

Lastly, since I work for the US Government and the Food and Drug Administration, I must include the following disclaimer: 'The opinions expressed here are those of the author and do not represent the official position of the US Food and Drug Administration or the US Government.'

1
The Welfare of Animals and Its Relevance to Our Health

> We must fight against the unconscious cruelty with which we treat animals...Until he extends the circle of compassion to all living things, man will not himself find peace.
> —Dr Albert Schweitzer

Our common fate

Whether humans will ever find peace is up for conjecture, but this book aims to demonstrate that until we improve the welfare of non-human animals, we will never find health. For many involved in the health field, this proclamation will come as a great surprise. For others, it might be viewed as approaching heresy. How could the medical field, which is charged with the enormous responsibility of promoting human health and alleviating our suffering, also be concerned about the welfare of animals? It may be argued that animal welfare has nothing to do with human health, or even, more broadly, with human welfare. Yet, the notion that the way in which we treat animals impacts our own welfare is not a new one. Philosophers, scientists and other thinkers, dating from ancient Greece to modern times, have long suggested that when we disregard the welfare of other animals it may come back to haunt us in one way or another. The list of such thinkers is long and includes distinguished names such as Pythagoras, Plutarch, Socrates, Albert Einstein, St Francis of Assisi, Jean Jacques Rousseau, Immanuel Kant and Isaac Bashevis Singer.

Much of these earlier reflections were speculation, but today there is mounting evidence of a very real, and often very direct, relationship between animal welfare and human welfare, most specifically with regard to human health. While this book will focus mostly on human

1

health, other aspects of human welfare (such as freedom from violence, crime and hunger) that have been connected with human health and animal welfare will also be explored. For instance, as shown by the opening quote, Schweitzer suggested that the poor treatment of animals can result in the inability of humans to find peace. Moreover, a connection has been made between cruelty toward animals and violence toward humans. This book will demonstrate how our health can be greatly influenced, positively or negatively, by how we choose to treat animals. By taking a look at some recent, highly publicized events that have threatened human health and welfare, a better picture of how human health and animal welfare are connected can be formed.

For instance, Chapter 4 will describe how, in 2003, avian influenza (H5N1 strain) spread swiftly across poultry farms in Asia and jumped the species barrier to infect humans, raising red flags that the next pandemic could originate from animal farms.[1] When, in 2009, swine (H1N1 strain) influenza swept across the globe, it was confirmed that animal agriculture can play a significant role in the emergence of new strains of influenza viruses. Animals, living in profoundly filthy and crowded conditions that severely compromise their welfare and their immune systems, are now predominantly raised for food in 'factory farms' or confined animal-feeding operation. This creates perfect breeding grounds for new infectious diseases that are potentially deadlier than those already witnessed.

Another way in which human health and animal welfare are connected can be seen in the responses of people during Hurricane Katrina in 2005. The world watched live television as many Louisiana residents refused to evacuate their homes in the wake of the hurricane and, in some cases, risked death to avoid losing their companion animals (who were not permitted on Coast Guard rescue vehicles or welcome in local shelters). Some 44 percent of those who refused to evacuate did so because they did not want to leave their animals behind.[2] Indeed, the most common reason people return to evacuation sites is to rescue their pets.[3] Post-Katrina studies show that the loss of these companions worsened the mental trauma many people experienced.[4] This was a wake-up call for public health and rescue agencies throughout the world to take the human–animal bond seriously and incorporate animal rescue into emergency plans, since not doing so puts human health and welfare at risk (not to mention the health and welfare of the companion animals in question).

In 2007, a worldwide recall of pet food that had been exported from China and was contaminated with melamine was ordered after possibly thousands of animals fell ill or died.[5] In addition to the public outrage this prompted, there was a perceived human risk since some of the tainted pet food was also fed to animals later processed into human food. Because of this incident, the need for improved regulatory monitoring of food fed to animals, so as to protect both humans and animal companions, became evident.[6]

In 2009, a woman in Connecticut, USA, suffered a dreadful attack by a pet chimpanzee; and in 2010, SeaWorld animal trainer Dawn Brancheau was killed by an orca in front of a horrified audience. These events underscored the dangers of using wild animals for entertainment and as pets.[7] No one knows for certain why these animals attacked but, as explored in Chapter 3, wild animals raised as pets or used for entertainment are often kept in deplorable or inadequate housing conditions, deprived of any semblance of a natural life and may be subjected to other forms of abuse. Although a direct link between these conditions and such attacks can be difficult to establish, having wild animals in close proximity to humans certainly increases the opportunities for such attacks and can expose people to novel infectious diseases, posing immense public health risks.

Arguably, one of the most significant and pressing public health issues of our time is climate change. Reports from the United Nations' Food and Agriculture Organization in 2006 and the Pew Charitable Trust in 2008 have increased awareness and acknowledgment of the connection between what we eat and climate change.[8] Chapter 5 describes how the unprecedented worldwide demand for meat and the subsequent rise in factory farming affect our climate in significant ways. Modern animal agricultural practices contribute more to greenhouse gas emissions and environmental degradation than many of the industries on which we have traditionally pinned the blame.

These examples all illustrate the important connection between human health and animal welfare. As will be demonstrated throughout this book, when we treat other animals well, there are clear direct and indirect benefits to human health. When we treat animals poorly, our collective health suffers. Chapter 2 will demonstrate how the abuse of animals is correlated with the abuse of humans, particularly women and children. In Chapter 3 we will explore how our shipment of animals around the globe for food, fur, the purposes of entertainment and the acquisition of exotic pets is directly and indirectly linked to some of the most dangerous epidemics we have faced in recent decades. Chapters 4 and 5 explore how factory farms are polluting our land, water and air and are making us ill. Chapter 6 presents mounting evidence demonstrating that the use of animals in experimentation is unreliable, inefficient and dangerous to human health.

On the positive side, as Chapter 6 further explores, when we strive to improve the welfare of animals by reducing their numbers in experimentation and replacing their use with non-animal research methods, we discover testing methods that more successfully predict human reactions to toxins and drugs. These human-based testing methods are proving to be better at protecting people from harmful chemicals than animal experiments, as well as providing the answers we need to find effective treatments and cures for the diseases that ail us. As Chapters 4 and 5 show, reducing our consumption of animals would confer great personal health benefits by substantially decreasing our risks of strokes, diabetes and cancers, and by helping us live longer, healthier lives. Additionally, it would decrease the number of factory farms and thus the number of animals living in such conditions. This in turn would help us avoid pandemics and limit further destruction of our environment.

Despite the link between animal welfare and human health, animal welfare issues have been, with few exceptions, notoriously absent from the public health dialog. When the subject of animals does enter discourse on human health, it is usually to highlight how animals are sources of infection for and cause injuries to humans. There has been almost no discourse, however, about the fact that the *way in which we treat other animals* is often central to how and why humans are injured or catch infections. Moreover, such treatment is central to how and why we face a significant number of health threats. This distinction is important to note. To illustrate this, it helps to take a look at how the public health discipline has traditionally approached the subject of zoonoses—infectious diseases that can be transmitted from non-human animals to humans (and vice versa). There is ample medical literature exploring the variety of zoonoses that can be passed from animals to humans. There is also much medical literature and many studies exploring how and why we contract zoonoses and pass them on to each other. Such work looks predominantly at human–human interactions, weather patterns and the life cycle of infectious agents. When public health practitioners and investigators do turn their attention specifically to animals, it is usually to describe which animals are likely to be infected and thus pose a threat to us. Rarely do we explore deeper than this and ponder whether the nature of our relationships with animals could play a role in whether they become infected with a pathogen in the first place or pass the pathogen on to us.

When human infection from a zoonotic disease grows to epidemic proportions, the public health response is frequently swift and unforgiving: we simply kill or 'cull' animals in hopes of thwarting a pandemic. When, in 2003, severe acute respiratory syndrome (SARS) swept across the globe, civet cats sold at wet markets (open markets selling live animals) in China were found to be infected with the SARS coronavirus. Civets are one of the thousands of species bred in captivity or captured from their natural habitats to be sold for exotic meat, fur, traditional medicines and other commercial purposes in the massive worldwide wildlife trade. Upon discovering the SARS coronavirus in civets, China's public health response was to kill thousands of these and other animals by drowning them in disinfectants and electrocuting and poisoning them.[9] In addition to killing, other tactics involved dosing animals with massive amounts of vaccines or antimicrobials, which creates the potential to cause additional problems, such as antimicrobial resistance.

After China's mass execution of civets, research later revealed that bats, not civets, are the likely primary source of SARS. But we should not rush to a judgment on bats either. Bats, as well as civet cats, are victims of the wildlife trade, which, as argued in Chapter 3, is the true culprit. As explored there, both the wildlife trade and wet markets are guilty of deplorable and inhumane conditions in which animals are kept and, frequently, killed. Because of these conditions, animals in the wildlife trade easily experience compromised immune systems, which render them highly susceptible to contracting pathogens from other animals. Evidence points to the wildlife trade and the live markets as being the nexus in which the SARS coronavirus passed from bats to other animals, including civets and, ultimately, humans.

Undeniably, pathogens transferred from animals to humans have led to human epidemics, but rarely do we ask whether the way we treat animals could have played a role in the epidemics, and whether better treatment might have prevented them from beginning in the first place.[10] By ignoring the welfare of animals, we fail to see the most direct and comprehensive (and sometimes the simplest) solutions. In the case of SARS, such an epidemic might never have occurred if we had treated animals better. If we had considered the welfare of animals, we would likely have placed greater restrictions on their sale in wet markets and for the wildlife trade. The reduction in, or elimination of, these practices would have reduced our chances of contracting a pathogen from animals. At the very least, if we had improved their conditions, perhaps fewer animals would have contracted SARS and passed it on to us.

Undoubtedly, treating other animals better is not the solution to every public health problem we face and, of course, no matter how well we treat animals, they may still become infected and end up being associated with the next major public health calamity. But we are being remiss, and arguably grossly negligent, when we avoid discussing our treatment of other animals as a potential cause and potential solution altogether. A considerable amount of human suffering may be avoided if, rather than asking ourselves how to thwart an epidemic once it has begun, we instead ask ourselves whether we can prevent an epidemic by treating animals differently.

The consideration that our treatment of non-human animals impacts human health and disease notwithstanding, there are several objections that can be raised when considering the welfare of animals and its link with human health. The first potential argument against considering animal welfare is that it makes no sense to discuss their welfare because, in particular, animals don't feel pain or suffer (there are other potential components to their welfare, such as living in freedom or being among their own kind, which are touched upon briefly in this book). Thus, as this argument goes, because animals are incapable of feeling pain and of suffering, their welfare is a non-issue. Alternatively, one may acknowledge that animals can and do feel pain and perhaps suffer, but that their potential for these is extremely limited or fundamentally different from that of humans. Again, with respect to pain and suffering, some may argue that the welfare of animals is a non-issue or, at best, a minor issue. The second objection is, assuming that one acknowledges that animals can feel pain and suffer, that human health nevertheless takes priority over the welfare of animals. The third objection, an extension of the second, is that animals must be utilized in ways that may cause their suffering in order to promote human health. Finally, the last major objection is that animal welfare is a social issue that is outside the purview of public health. Each of these objections is addressed in the following sections.

Pain and suffering

Do animals feel pain and suffer? It is very tempting to bypass this question and assume that the reader, as most people, including the general public and scientific experts, acknowledges that all mammals and most vertebrate animals can experience pain and can suffer. However, there are still those who argue that animals cannot experience pain and/or suffering and who use that assertion as a justification to deny the legitimacy of animal welfare and defend some of our current treatment of animals.

The terms 'pain' and 'suffering' are often used interchangeably, but there is widespread debate as to what suffering actually means and discussion of how certain forms of suffering may be distinct from the experience of pain. Although this is a complex topic that others have written extensively about, a few of the more relevant points can be addressed here. First is the question of pain. Pain is usually taken to mean a negative sensation in response to adverse or noxious stimuli. The majority of people today, including philosophers, animal behaviorists, physiologists and other scientists, acknowledge that vertebrate animals feel pain. In addition, substantial scientific evidence demonstrates that most, if not all, vertebrates, including all mammals, birds, reptiles and fish, as well as many invertebrates, such as cephalopods and crustaceans, feel pain.[11]

Vertebrate animals display nearly all the same external signs to an unpleasant stimulus as humans: they writhe, cry out (some within our own range of sounds and some ultrasonically), yelp, show facial contortions, avoid putting weight on an injured limb, demonstrate decreased appetite and interrupt their normal behaviors. These external signs distinguish most animals from other organisms, such as an amoeba, which may avoid a harmful stimulus but demonstrate no other manifestation that suggests an experience of pain. Animals also show characteristic physiologic reactions suggestive of pain. Like humans, they show elevated respiration and heart rate, dilated pupils, release of stress hormones (e.g. cortisol and norepinephrine), and they often respond to pain-alleviating medications in a similar fashion to humans. Even fish, when subjected to chemical irritants, demonstrate a response to morphine, suggesting that they experience pain and that the drug can alleviate it.[12] We can also infer animals' pain by the structural similarities of their nervous systems to our own. While inter-species differences exist, all vertebrates evolved to share many of the basic neurophysiological features that allow for the processing and perception of pain. For example, nociceptors (pain receptors) are widespread in the tissues of a wide range of animals, and chemical modulators of these receptors, such as endogenous opioids, exist in both humans and non-human vertebrates—and at least in some invertebrate animals. Thus, the weight of evidence points to animals (at least the animals discussed in this book) as being capable of experiencing pain. With respect to the topics covered here, since much of our treatment of animals involves causing them pain, animal welfare deserves serious attention.

The second question concerns the topic of suffering. There are many philosophers and scientists who maintain there are forms of suffering that are distinct from pain and, moreover, that there are many different concepts and criteria of suffering.[13] Pain can be one component of suffering, but it is not a necessary component. Common forms of suffering that do not necessarily involve pain include negative cognitive or psychological experiences, such as the experience of fear, anxiety, distress, emotional trauma, foreboding, loneliness and terror.[14] For example, if a woman experiences a stroke and becomes paralyzed on one side of her body, she may suffer from the stroke even if no pain is involved. She may suffer because she is unable to perform what were previously normal functions for her and she may experience distress and anxiety as a result of her limited function.

Pain can also cause limited function, which can lead to negative emotional or cognitive experiences. An individual with an injured leg may be limited in how he or she uses that leg because of the pain that may follow. This limited function may cause distress and immense frustration, which can be forms of suffering apart from the experience of pain itself. It has been argued that animals lack cognitive and emotional capabilities to experience distress, terror, anxiety, loneliness and the like, and thus cannot suffer in these ways, or they suffer much less in comparison with humans. But is it true that animals lack these abilities? This next section will briefly present some of the amounting evidence that shows animals have a wide range of emotional and cognitive capabilities and are therefore likely to suffer—at least in terms of their ability to experience negative emotional and/or psychological states, such as distress, terror and loneliness.

The evidence that animals can suffer from negative emotional and cognitive states is accumulating, in part, because of a revolution in the field of animal behavior, which is breaking the long-held taboo against studying animal minds.[15] There is now an explosion of work by notable scientists, such as Frans de Waal, Michael Tomasello, Marc Bekoff, Sue Savage-Rumbaugh and Jaak Panksepp, examining the mental lives of animals (how they think and feel). These studies reveal that many animals experience a range of emotional and cognitive capacities that were previously denied to them. In what follows, a brief array of evidence from such studies will be presented to indicate these capacities. The evidence is such that it is extremely difficult to deny that animals are incapable of suffering, and it shifts the burden of proof from having to show that animals suffer to having to show that animals do not.

Language and cognitive abilities

Many have heard of the famous chimpanzee, Washoe, who was the first non-human to learn parts (albeit truncated) of American Sign Language. What might surprise some, however, is that she passed on a portion of her sign language knowledge to her adopted chimpanzee son, Loulis, without any inducement from humans.[16] Stimulated by the success of Washoe's language abilities, other investigators achieved similar levels of communication with other chimpanzees, gorillas and orangutans.[17] Orangutans have learned to communicate with humans using computer touch screens.[18] The communication skills of the great apes have many scientists arguing that they do indeed possess the capacity for some form of language, the primary attribute that some believe separates humans from all other animals.[19] Chimpanzees also show evidence of spontaneous acts of kindness, social awareness and culture.[20] From such evidence, some have argued that they exhibit 'theory of mind'—that is, the ability to attribute beliefs, desires and other reference-making abilities to other beings, be they human or chimpanzee, and that they rely on such attributions to inform and shape their interactions.[21]

Tufted capuchin monkeys, a New World primate, recognize photographs of faces of familiar monkeys and can pick out photos of individuals who do not belong to their social groups.[22] Studies suggest that rhesus monkeys know when they know something and know when they don't. In a series of experiments, researchers found that these monkeys chose to take only memory tests on which they were found to perform well and declined to take tests they would likely fail.[23] This suggests that the monkeys could gauge the apparent difficulty of a test or knew whether or not they remembered the correct responses. This ability to make adaptive decisions about future behavior contingent on the current availability of knowledge is a capacity associated with cognition in humans.[24]

Emotional and other abilities

Given their close genetic relationship with us, it's not surprising to discover the nuanced cognitive traits in non-human primates. But what might surprise many is that even the 'lowliest' of animals—the ones who throughout history we have most disregarded—are far more complex than we commonly believe. To take an example, studies suggest that rats and mice experience empathy. In one famous, yet lamentable, experiment performed in the 1950s, psychologist Russell Church trained rats to obtain food by pressing a lever.[25] He found that if these rats saw that, by pressing the lever, another rat in a neighboring cage would receive an electric shock, the first rat would interrupt her activity. A more recent and highly publicized (and also not so nice) study reveals a potential form of empathy in mice.[26] Here, researchers injected painful irritants into mice, then watched them writhe in pain. When other mice witnessed this, they showed heightened sensitivity to pain themselves. But the observer mice mostly only reacted to the pain being experienced by other mice with whom they were familiar. This strongly suggests that the responses of the mice were more than simply an automatic fear reaction and that they were selective to specific mice—a form of empathy. In a review of empathy studies in mice and rats, leading animal behaviorist Frans de Waal stated,

> Apart from a few rear-guard behaviorists, few people hesitate to ascribe empathy to their dogs. But then dog is man's best friend, freely credited with lots of human sentiments. You wouldn't expect a hard-nosed scientist to make similar claims about, say, rodents, would you? Yet a significant line of research . . . demonstrates not just empathy's existence in rodents and other animals but its subtleties and exceptions as well.[27]

Tickle a rat's belly and he will emit ultrasonic chirping sounds, which are believed to have the same neural basis as human laughter.[28]

In the 1980s Gerald Wilkinson made a significant discovery about vampire bats: they are altruistic. These bats feed on the blood of larger mammals. Food sources can be scarce so that they may go for long periods of time without food. If a bat is starving, other bats will regurgitate blood to share with him or her, compromising their own nourishment.[29] Hens display nuanced communication with other hens.[30] For example, they use specific calls to relay detailed information about food to other hens.[31] They finesse their calls when relaying to others that a found food is particularly preferable. Chickens demonstrate the ability to recognize each other by their facial features.[32] Several studies also suggest that they have the ability to understand that an object, when taken away and hidden, continues to exist—a feat beyond the capacity of very young children.[33] Jay birds use deceptive tactics to hide food when they know other birds are watching them.[34] What is perhaps more interesting, however, is that such deception most frequently occurs when the jay birds hiding the food have previously stolen food from other birds. This suggests that they make assumptions about the actions of other birds based on their own. In other words, it takes a thief to know one. Scrub jays also display memory, and an awareness of the passage of time and its effect on different foods. Pigs can show a pessimistic or optimistic outlook on life, depending on whether their environment is dull or stimulating.[35] And a recent study found that domestic pigs quickly learn how mirrors work and will use their understanding of reflected images to scope out their surroundings and find their food.[36]

Reptiles are commonly thought of as being unemotional creatures. However, physiologist Michael Cabanac demonstrated an emotional fever response in lizards and turtles resulting from human handling.[37] This is akin to studies which demonstrate that in rats body temperatures rose by 1 °C or more when handled by unfamiliar persons. This response disappears when the same person handles the rat repeatedly over several days, as the rat develops trust in the handler. If another handler is introduced, the emotion-caused fever returns. The similarity in the fever response between rats and reptiles suggests that reptiles also experience fear or distress, which can manifest as a rise in body temperature.

Continuity with humans

These are just a few of the many examples of studies revealing animal emotion and cognition. Given the evolutionary and biological continuity across species, it is more and more difficult to claim that there exists any strong dichotomy between humans and other animals, at least when it comes to basic emotional and cognitive abilities that can give rise to suffering.[38] Eminent psychologist and Senior Fellow in Theoretical Neurosciences at the Neurosciences Institute in California, Bernard Baars, summed up the evidence rather nicely:

> All animals engage in purposeful action … other animals investigate novel and interesting stimuli as we do, ignore old and uninteresting events just as we do … Human beings are different of course … But we can no longer pose absolute barriers between ourselves and other animals … [39]

Granted, many of the studies on animal minds are preliminary and there are other studies that may contradict some of the above findings. Far more needs to be done to better delineate which capabilities exist and in which species. At the very least, however, these initial studies suggest that some emotional and cognitive capabilities previously attributed only to humans are more widespread than previously thought, even if many of these capabilities may be restricted in other species as compared with humans. And, as we devise more sophisticated (and kinder) ways to study animal minds in their natural environments and contexts, we will likely discover that they posses many more capacities than we currently assume.[40] As Frans de Waal stated, 'efforts to single out distinctly human capacities have rarely held up to scientific scrutiny for more than a decade, such as claims about culture, imitation, planning, and the ability to adopt another's point of view'.[41] Even aquatic animals, such as whales and dolphins, who are quite removed from humans on an evolutionary scale have surprised us with their abilities. Carl Sagan's famous tongue-in-cheek quote is a humorous reminder of some of the capabilities which exist in such animals: 'while some dolphins are reported to have learned English—up to fifty words used in correct context—no human being has been reported to have learned dolphinese'.

The notion that we may share some basic cognitive capabilities with other animals makes evolutionary and biological sense.[42] As the scientific evidence demonstrates, animals are not mere automata, living solely off instinct; they display every indication of being able to feel pain and of having emotional and cognitive abilities that rely on more than mere stimulus–response behaviors. In other words, science shows that animals have mental lives and that their actions are purposeful.

In fact, our very use of animals in experimentation is often predicated on these abilities. According to Colorado State University bioethicist and Professor of Animal Sciences Bernard Rollin, 'not only does much scientific activity presuppose animal pain as we have seen vis-à-vis pain research and psychological research, it fits better with neurophysiology and evolutionary theory to believe that animals have mental experiences than to deny it'.[43] For instance, many of the medical experiments conducted on animals to understand human disease, particularly psychological experiments, reveal just how much science as an institution agrees that animals experience pain and have various cognitive and emotional abilities. Scientists have intentionally caused and studied the effects of chronic depression, post-traumatic stress disorder, obsessive-compulsive disorder, severe anxiety, schizophrenia and dementia in animals, including mice and rats. The scientific community often attempts to distinguish what animals experience from human experiences by referring to the psychological states induced in the laboratory in a type of scientific doublespeak. For example, animals in laboratories aren't routinely described in the biomedical literature as experiencing depression but as displaying depression-like signs.

Yet, if the scientific community did not believe that animals were capable of experiencing these complex psychological states and conditions, why would these studies be conducted in the first place? If the scientific community did not believe that animals experience depression, why would studies on drugs intended to treat depression be conducted on animals? If the scientific community did not believe that animals possess at least some cognitive abilities, why would studies on the *loss* of cognitive abilities (for research in dementia) be conducted in animals? It stands to reason that the scientific community operates on the assumption that animals possess a wide range of emotional and cognitive capacities. And it makes sense to believe this: the animals in these studies display signs remarkably similar to those seen in human forms of these psychological illnesses or conditions. If these animals exhibit these signs in fairly systematic ways, then, in conjunction with their neurophysiological similarities and evolutionary continuity, it is a natural and scientifically sound inference to conclude that they experience chronic depression and other emotional and psychological states. In other words, animals don't just show distress, they *feel* it.

Further, if these animals experience depression, anxiety and distress, then it is plausible that they are also suffering in these circumstances. These experiences certainly count as suffering in humans and there is not any scientifically sound reason to believe that this is not also the case in animals. In fact, the US Public Health Service Policy on Humane Care and Use of Laboratory Animals reflects the implicit scientific acceptance of these psychological states in animals and their ability to suffer by suggesting that researchers 'limit the use of animals or limit animal distress' and that 'investigators should consider that procedures that cause pain or distress in humans beings may cause pain or distress in other animals'.[44]

There will be those who will take what has just been stated as evidence that animals are effective tools for medical research. They may contend that if humans and animals share some of the same basic neurophysiological processes and cognitive and emotional abilities, then this implies that animals make good models for human disease. But this argument would be missing a vital element. Medicine now deals with the subtle nuances of physiological mechanisms in order to precisely target an intervention, such as a drug to boost or inhibit a specific cellular or genetic process. The more we study other animals, the more we learn how species—and even strains within a species—differ in these subtle mechanisms. It is specifically these differences that render the reliability of animal experiments to predict human outcomes highly questionable. This topic is further explored in Chapter 6.

Summary of suffering

As animal behaviorist Marc Bekoff states, 'we must be careful neither to imbue animals with unknown cognitive capacities nor to rob them of skills they might possess'.[45] No doubt species differ in their cognitive and emotional capabilities and there are forms of suffering that are likely to be unique to humans. A dog may not suffer from knowledge that he may not meet his grandchildren, though a human might. But even if animals sometimes suffer differently from humans, why should that matter? An animal's experience does not need to be identical to a human's for it to count. Consider color perception as an example. Goldfish, Japanese dace fish, turtles and many birds are tetrachromats—that is, their retinas contain four classes of cone cells for picking up information from wavelengths of light in the red, green, blue and ultraviolet parts of the color spectrum.[46] There is fairly good evidence that pigeons and ducks may even be pentachromats (possessing five channels or cone cells for conveying color information).[47]

In contrast, normal humans are trichromats, with three classes of cone cells picking up information from the red, green and blue parts of the color spectrum. And about 1 percent of men are dichromats, only having the ability to pick up information from blue and either red or green wavelengths of light. Because of this difference, tetrachromats can rely on a broader range of information to see the world than we do. Tetrachromatic hummingbirds, for instance, can distinguish near-ultraviolet light (370 nm) both from what appears to us as darkness and from 'white' light lacking wavelengths below 400 nm.[48] Humans cannot perform either of these tasks. Does the fact that humans perceive color differently, and in a more limited way, than hummingbirds imply that humans don't see color? Of course not; we just process color information differently. The same analogy can be applied to the question of animal cognition and the capacity of animals to suffer. Even if animals do not suffer in all the same ways as humans, this does not mean they don't suffer. We may also want to acknowledge that there may be forms of suffering that are unique to other species. If we destroyed a hummingbird's ability to perceive ultraviolet light, this might lead to a form of suffering unimaginable to humans. By destroying a hummingbird's ultraviolet perception, a function that is normal for her, we might be causing that bird to suffer even if we did nothing else to her.

Let's suppose, for the sake of argument, that all other animals lack the cognitive sophistication we ascribe to humans (whatever that sophistication may be) and, in particular, the capacity for rationality. Thus, as one argument goes, because animals lack rational ability, their capacity to suffer is greatly reduced or perhaps non-existent. If we granted this argument, we would have to admit that humans lacking rational ability, such as the severely mentally incapacitated and infants, would also have greatly reduced capacities for suffering, and perhaps these would be non-existent. It is indeed likely that certain forms of suffering are reduced in non-rational beings, be they certain humans or non-humans. For example, a rational being may suffer from fear of losing his job and the inability to provide financial stability for his family. But just as there are forms of suffering that may be greater in rational beings, there may also be forms of suffering that are greater in non-rational beings. A non-rational being cannot rationalize or explain away his suffering.[49] Consider that an animal captured from the wild does not comprehend what is happening to him and may experience raw terror, even if he is just being implanted with a tracking device for his protection prior to being released. In such cases, animals may experience far more terror and suffering than humans would.

The lack of comprehension and inability to intellectualize their experiences is what often makes animals, not unlike infants and severely mentally incapacitated humans, more vulnerable to certain forms of suffering. Additionally, the pain and/or suffering experienced by a non-rational being may take up the whole of his experience. A common therapeutic method employed to treat prolonged mental suffering or pain in people is cognitive behavioral therapy. This teaches people to override or mitigate their negative experiences by altering their thought patterns and their behavioral responses. Thus humans can be taught to use thought to escape pain and other forms of suffering. The intensity of pain can also be mitigated through expectations, memories and the consideration of, and attention to, other interests and situations.[50] A non-rational being, on the other hand, may not be able to mentally escape or expect the pain or other negative experience to end and his whole attention may be on that negative experience.[51] A baby crying out in pain is a tragic thing to most of us precisely because we suspect that his whole awareness and attention at that moment is on his pain and, as a result, he may be suffering tremendously. Similarly, an animal's pain or negative emotional experience (such as terror) may be magnified because of the lack of ability to mentally escape it or rationalize it; as a result, that animal may also be suffering tremendously.

Whatever we determine about the cognitive abilities of animals, the presence or absence of certain forms of cognition may not preclude the ability to suffer but only, perhaps, prevent (or conversely increase) the ability to experience certain forms of suffering. Though other animals may lack the complexity of many of the emotional or cognitive capacities possessed by adult humans, this does not imply that their experiences are any less relevant to them. To paraphrase Mark Bekoff, even if a mouse does not have the same sense of 'I-ness' as you or I do, this does not mean that he cannot feel that something is painful or pleasurable to him and cannot suffer or enjoy.[52]

Ultimately, what is relevant here is not how one suffers but the fact that one does suffer. There are indeed differences between humans and other animals, but these differences don't in themselves deny the presence of suffering in animals. The accumulation of evidence certainly demonstrates that animals can feel pain and the evidence is mounting that animals possess a variety of emotional and cognitive abilities. Coupled with their biological and evolutionary continuity with humans, this evidence forms a reasonable basis for assuming that animals can suffer from negative emotional and psychological states, such as distress and fear. Given the evidence, and given the immense harm we may be causing animals by denying or ignoring their suffering, the burden of proof should be on those who maintain that animals do not suffer. If a chicken lives her entire life in a cage so small that she is unable to stretch her wings, an activity that would be normal, the burden of proof should be on those who maintain that we have not caused her to suffer. If there is a possibility that a cat in a laboratory lives in daily terror of what might be done next to him, the burden of proof should be on those who argue that he is not suffering. When taken as a whole, the aforementioned evidence offers a strong case for the existence of animal pain and suffering, and this book will proceed on the premise that the animals discussed experience both and that, therefore, their welfare is relevant.

Other potential objections

Even if human health advocates acknowledge the ability of animals to suffer, it may still be objected that human health and welfare take priority over animal welfare. This brings us to the second and third main objections that may be argued against including animal welfare in public health. There is a common misconception among health professionals that the maximization of human health requires a disregard for, or the minimization of, the welfare of other animals. In other words, there is a perceived conflict between human health and welfare and the welfare of animals. An extension of this position is that we depend on using animals in ways that require that we disregard or minimize their welfare in order to promote human health. That is, animal welfare is not only irrelevant but in opposition to human health. For example, some may argue that factory farming is the most efficient and economical method of producing enough food for and preventing undernourishment in the growing human population. But, as is argued in Chapters 4 and 5, not only does factory farming threaten our health by contaminating our environment and generating infectious diseases, we can actually feed more people by reducing our use of animals as food.

Perhaps the example most frequently cited in support of minimizing the welfare of animals is animal experimentation. Experimenting on animals is a highly controversial issue. Advocates of this practice frequently employ the fallacy of a false dilemma: that we must choose to care about human suffering or about animal suffering, and that we cannot do both. This erroneous thinking leads us to believe that we must either experiment on a mouse (or a dog or a monkey...) or we must experiment on a human child, implying that we are forced to make a choice—it's the animals or us. A hypothetical moral question frequently posed to advocates of animal protection is the sinking boat scenario. If you were on a sinking boat with a child and a dog, the scenario asks, and you could save only one, which would you save: the dog or the child? While this seems like an interesting moral dilemma (and one that is frequently used to force the reader to weigh the worth of other individuals, such as their spouse or their child), public health does not have to choose between the dog and the child.

Public health can cheat this scenario altogether. We generally have a third option: one that provides a way out of this polarizing quagmire, in which an advocate for humans can also be an advocate for animals, and in which these two roles are complementary. Although there will always be times when the welfare of an animal will be pitted against the welfare of a human (or humans), how we decide to resolve such issues should be done in a manner that causes the least amount of harm possible to all involved, human and animal alike. As this book will demonstrate, however, in most cases where humans interact with animals, not only can we help people without harming animals, we can also best help people by not harming animals. We will explore numerous examples illustrating that our treatment of animals is integral to many of the causes of, and potential solutions to, some of the biggest human health threats. In other words, not only can we save the dog and the child, but in order to save the child *we have to save the dog*. Our mental and physical health, the state of our environment, the safety of our food, and the efficacy and safety of our medical treatments are inextricably tied to how we choose to treat other animals.

A very brief history of public health

The final potential significant objection to the premise of this text is that issues of welfare, protection and ethics concerning our use of, and interaction with, animals are outside the purview of public health and medicine. Sociologists, animal advocates, philosophers and others, it could be argued, are better situated to deal with such social issues. The idea of including the welfare of animals in the sphere of public health may be seen as a radical one. However, public health is no stranger to social concerns (or radical ideas). On the contrary, public health, throughout its history, has been an integral part of social change. A brief look at the evolution of public health over the past two centuries reveals its expanding role in social issues.

Public health has frequently challenged, directly and indirectly, cultural mores and societal prejudices. At a time when illnesses were often blamed on the lifestyles of the poor, Dr John Snow and clergyman Henry Whitehead marched into the slums of London and identified the root of a merciless cholera epidemic sweeping across the city.[53] They proved that the cause of the 1854 epidemic was not the 'moral failings' of the lower classes, which was thought to make them more susceptible to cholera, but water contaminated by sewage.[54] The idea that contaminated water could be the source of an epidemic was a radical one at the time and was contrary to the predominant 'miasma theory', which contended that spreading of foul-smelling odors, mostly by the city's laborers, was the root cause.[55] Once the connection between sanitation and illnesses such as cholera was revealed, sanitation campaigns dominated public health practice in nineteenth-century Europe and North America. It became evident to health professionals that infectious diseases in the Industrial Age were rooted not only in increased industrialization but also in impoverished living conditions and inequities in social standing.[56] The 1854 cholera epidemic spread rapidly throughout London's labor population because of their dismal and crowded living conditions. The sanitation campaigns reflected an awareness of the powerful connection between people's social position, their living conditions and their health outcomes.[57]

German pathologist Rudolph Virchow, credited as being the 'father' of modern pathology and who greatly advanced public health, asked, 'Do we not always find diseases of the populace traceable to defects in society?'[58] Understanding the connection between social standing and health, he became a social justice activist on behalf of the poor.[59] It became widely acknowledged that infectious diseases could not be successfully combated until the underlying social causes were unraveled and tackled, and the sanitation campaigns transformed health care professionals into social activists.[60]

Since the sanitation campaigns, as new health concerns arose, the social roots of illness and health became increasingly evident and activism by health professionals grew dramatically. During the early twentieth century, public health expanded to address maternal and infant mortality and child undernourishment.[61] High rates of occupational disease and industrial injuries led to programs for industrial hygiene and occupational health.[62] Work-related health problems, such as coal workers' black lung disease and silicosis, led to campaigns for safer work environments.[63] Advocacy on behalf of women, children and workers for basic rights and protections, such as improved working conditions, limitations on working hours and restriction or banning of child labor practices, was an important extension of public health and medical practice.

Many social revolutionaries in public health emerged. Nurse Margaret Sanger was a rebel in her time. In the early twentieth century, she witnessed countless women dying during childbirth from unwanted pregnancies.[64] She challenged the long-held doctrine that prevention of conception was 'indecent and disgusting'.[65] Sanger became a fugitive of the law when she responded to the needs of destitute mothers by opening family planning clinics for the poor in New York City and advocating for women's rights to access to contraception and control over their own fertility.[66]

Since Sanger's time, women's and children's rights campaigns have led to the prioritization of child health care, improved reproductive health care for women, access to family planning and the emergence of public health programs, such as the Women's, Infants, and Children food service. By the latter half of the twentieth century, increasing awareness within psychiatric and patient communities resulted in collaborative efforts to tackle long-held stigmas against the mentally ill.[67] As these stigmas began to diminish, the promotion of mental health, in addition to physical health, was incorporated for the first time into routine public health policy. In 1946, the World Health Organization (WHO) astutely defined health as 'a state of complete physical, mental and social well-being and not merely the absence of disease or infirmity'.[68] Since its inception, WHO had also recognized the link between health and social structure, justice and social disparities. The founders shared a vision of health as pre-eminently shaped by social conditions.[69]

Over the course of the twentieth century, public health steadily increased its scope. Evolving needs and societal values led to substantial changes and the inclusion of ethics as a field in public health. Following the revelation of Nazi experimentation on war camp prisoners, and the aftermath of the infamous Tuskegee experiments in which African-American men with syphilis were denied access to penicillin in order for researchers to study the course of the illness, significant regulations were created for the protection of human subjects in research.[70] Ethics became an integral discipline within public health throughout the world.

The HIV/AIDS pandemic that began in the 1980s is arguably the single biggest health crisis to transform public health policy and practice.[71] The pandemic drew attention to the plight of some of the most vulnerable groups and the need to prevent discrimination in law and health practice.[72] When much of the world vilified homosexuals, public health workers zeroed in on this population as being in need of priority protection. The marginalization of and discrimination against gays, lesbians and bisexuals, the stigma that followed those infected with HIV and the disregard for their welfare led to a rather revolutionary approach to health. Jonathan Mann, then head of WHO's global AIDS program, led the call to acknowledge the vital linkage between health and human rights.[73] He recognized that HIV infection rates were closely connected to inequality, injustice, discrimination and the failure of public health to recognize the roots of vulnerability worldwide.[74] Through collaborative efforts with human rights activists, public health eventually recognized that the fundamental protection of human rights is one of the most effective means to ensure positive conditions for health. Public health also acknowledged that there is a moral obligation to protect and become advocates on behalf of the most vulnerable populations throughout its policies and practice.

As these examples over the past two centuries demonstrate, public health has come to recognize how social factors influence our health and, as societal values have changed, so too has the practice of public health. Public health now encompasses a vast scope of social and non-social factors, which were previously not recognized as having an impact on health. Ethics has become an integral arm of public health. Topics such as women's autonomy and reproductive rights in religious fundamentalist countries, health disparities in developed nations, poverty and disadvantaged populations in developing nations, the equitable distribution of medications and disparate burdens placed on minority groups now predominate public health discussions. Over the past two centuries public health has been challenged to go 'beyond its traditional work to embrace programs to decrease poverty, illiteracy, environmental degradation, and violence'.[75] Individual rights; discrimination based on gender, ethnicity, sexual orientation and age; political, legal and economic forces; poverty, war and violence; personal behaviors; the structure of urban environments and numerous other factors impact our health.

Alvin Tarlov, Director of the Texas Institute for Society and Health, explained that 'there are five major categories of influence on health: genes and associated biology; health behaviors, such as dietary habits, tobacco, alcohol and drug use, and physical fitness; medical care and public health services; the ecology of all living things; and social and societal characteristics'.[76] As stated by Stephen Leeder from the Menzies Centre for Health Policy in New South Wales, Australia, 'the new public health makes considerable reference to sociological and anthropological insights and engages with the world of human behavior... in pursuit of better health'.[77] As a result of these insights, physicians, nurses and public health practitioners have become an integral part and are, at times, at the forefront of social change. We now understand that how we interact with and treat each other, how we view others, how we share (or do not share) our resources, how we eat, how we work, how we play, how we shelter ourselves, how we think, how we govern ourselves, how we spend money and how we relate to our environment—in short, how we *live*—influence our health. Despite all that we have come to understand, however, we have yet to fully appreciate one of the major categories described by Tarlov: the ecology of all living things. It is true that we study in great detail the life cycle of mosquitoes and are applying greater attention to environmental changes. But we have always overlooked (with the few exceptions described earlier), and continue to overlook, a significant facet of human existence that has prevailed since our beginning: our relationship with and treatment of other animals.

The elephant in the room, her welfare and why that matters to our health

Today, food security issues, limited health care resources, unsustainable consumption patterns, environmental degradation, bioterrorism, global warming, human population growth, obesity, novel infectious diseases, world hunger and violence are now emerging as the most urgent public health issues that we face.[78] The complexity and multifactorial roots of these issues necessitate a public health strategy that goes well beyond the health sector. We have started to do this. Due to recent and significant changes in our climate, for example, the fields of public health and medicine have begun to acknowledge that how we treat our planet affects our health. To fully tackle these urgent issues head on, we must also acknowledge that many of them are inevitably linked with how we treat other animals. It is overly simplistic and inaccurate to say that every human–animal encounter that is connected with human illness is a result of our actions or is avoidable. Yet, a significant proportion of these connections are, at least in part, a result, directly or indirectly, of our disregard for or minimization of the welfare of animals.

If public health is concerned about the public's health, then it must address a series of issues that it has so far largely avoided—and that affect the welfare of animals. If public health is concerned with how climate change endangers human health, it should also be concerned about factory farming's impact on global warming. If public health is concerned about the threat of new and deadly infectious diseases, it should also be concerned about the wildlife trade's potential to unleash a pandemic worse than HIV/AIDS. If public health is concerned about limited health care resources, it should also be concerned about how a significant portion of our health care dollars is diverted into animal experiments of dubious medical value. If public health is concerned about the alarming rise in antibiotic resistance, it should also be concerned about the ubiquitous feeding of antimicrobials to animals on industrial farms. If public health is concerned about violence, it should also be concerned about the connection between violence toward animals and violence toward people. If public health is concerned about the epidemic of obesity, it should also be concerned about obesity's connection to our unprecedented consumption of animal products. If public health is concerned about the safety of our food, it should also be concerned about factory farming's effect not just on the safety of animal food products but also on that of our fruits and vegetables. In short, if public health is concerned about the protection and promotion of human health, and if public health acknowledges that every other facet of human existence plays a role in our health, it must also acknowledge that how we relate to animals is a major determinant of our health. If public health is concerned about public health, we must turn our attention to the elephant (and every other animal) in the room.

Public health must now recognize that when we disregard the welfare of animals and cause them to suffer, we also disregard our own welfare and cause our own suffering. Today, humans are harming animals on a record scale. Worldwide, over 64 billion land animals are raised and slaughtered for food annually.[79] Most of them are now reared in massive factory farms, a source of immense animal suffering.[80] Annually, more than 115 million animals suffer as they are used in experiments or to supply the biomedical industry throughout the world.[81] Every year, billions of animals are ripped from their natural habitats and sold as pets and entertainers, or killed for their fur, skin, meat or other body parts. And, with our own population explosion, we are destroying habitats for other animals at a rate that might soon become irreversible.

Despite the rise in our maltreatment of certain animals, we are seeing an unprecedented increase in the number of humans sharing their lives with other animals solely for the purpose of companionship. This is a drastic change from much of human history in which our interactions with animals revolved mostly around our dependence on them for food, transportation, labor and similar uses. In the USA, for example, 62 percent of households now include companion animals.[82] Increasingly, couples, empty nesters and professionals are living with companion animals in lieu of having children.[83] Companion animals are now commonly viewed as integral family members. We see them as our kids. We lavish them with attention. We invest in their health and welfare. We mourn their deaths. And, we spend lots and lots of money on them. Americans now spend about $41 billion a year caring for companion animals. The companion animal industry is one of the fastest growing in the USA, and similar market increases are occurring in other parts of the world.[84]

We are increasingly attached to the animals with whom we share our lives but at the same time have become increasingly detached from other animals, such as those we consume for food. The result is our inadvertent (or deliberate) complicity in the suffering and slaughter of billions of animals each year. Our contradictory relationship with animals is nothing new. Historically, we have tended to view certain animals as 'pets' worthy of our regard and others as 'pests', whose welfare we have discounted. Animals with whom we are less familiar, who may be less aesthetically pleasing to us or who have been used as food or tools have always carried the brunt of our maltreatment. But what is different—and allows for unprecedented amounts of suffering—in more recent times is this: animal mistreatment is hidden from the public eye. What happens to animals in factory farms, in experimental laboratories, in circuses and other 'entertainment' venues, in fur farms, in individual homes and during transport across the globe usually takes place behind closed doors.

Slowly, this suffering is coming to light. Daily, news sources cover animal welfare issues and the public is increasingly expressing outrage against abuses of animals. Public perception of animals is starting to change—in part because of our growing relationship with and understanding of some animals, and in part because of the scientific evidence exploring animal minds. Animal protection has gained ground and is now a significant social movement. A wide range of academics and professionals, including philosophers, ethicists, legal and religious scholars, authors, sociologists and physicians, have joined the movement. As a result, laws are being passed worldwide to confer greater protection for animals in many circumstances. For instance, in 2009, voters in the state of California banned the use of battery cages for egg-laying hens and of veal and sow gestation crates, following the lead of four other states. The entire European Union (EU) will ban battery cages in 2012.[85] In 2010, the Spanish region of Catalonia made headline news by becoming the first mainland region of Spain to ban bullfighting, a centuries-old tradition. For the first time in its history, China, in 2009, unveiled legislation to address deliberate cruelty to animals.[86] Norwegian law has included fish in its Animal Welfare Act since 1974. Since 1997, The EU officially recognized animals as being sentient, able to feel pain and experience emotions, and this was written into their basic treatise. The legally binding protocol requires the EU and its member states to 'pay full regard to the welfare requirements of animals' as 'sentient beings'.[87] A record 91 animal protection laws were passed in the USA in 2008, eclipsing the previous record of 86 new laws in 2007.[88]

Discourse about the use of animals in experimentation and alternatives to their use is becoming of greater interest, and the practice of eating vegan diets—diets that exclude all animal products—is a growing worldwide phenomenon. There are around 1500 books on animal ethics today and at least 12 academic journals in the field of animal ethics.[89] In the USA, animal law is one of the fastest growing fields in legal scholarship.[90] The study of human–animal relations is a burgeoning interdisciplinary field in sociology, psychology and zoology on both sides of the Atlantic.[91] Worldwide, there is increasing recognition that animals have inherent value and there is growing skepticism over the long-standing belief that the suffering we cause them is morally irrelevant.

Public health must also recognize the welfare of animals. We must pay attention to whether or not the animals we encounter are abused, are forced to endure their lives in confinement, live under daily terror, are deprived of environmental stimulation, are removed from their own kind or are deprived of meaningful social interactions because all of these factors may affect humans. A beaten elephant may become a rampaging elephant; an abused cat may be a stepping stone for someone on their way to abusing a child; a distressed pig in a factory farm may be a reservoir of infectious diseases; and a terrified rat in the laboratory may produce study results that endanger human health.

The chapters that follow will explore these correlations in greater detail. First, it should be stated that, while humans are obviously also animals, most often the term 'animals' will be used rather than 'non-human animals' for convenience throughout this book. Second, rather than narrowly being defined by the strictest definition, the term 'public health' is here meant to include all health-related disciplines and professionals, including health care practitioners, public health officials, research scientists and medical ethicists. Third, this book serves as an introduction to animal welfare and public health. It is by no means meant to be an exhaustive account of how every human–non-human relation affects our health. Rather, it is meant to be an overview of some of the major ways in which we encounter and/or use other animals and how those interactions or uses affect human welfare.

Sadly, we have hardly begun to consider the thousands of ways that the human–animal relationship impacts human health. But once the public health profession recognizes the importance of this relationship, it will be able to start finding effective solutions for many of the causes of so much suffering and ill health in both humans and animals. Fortunately, many individual health practitioners are starting to recognize the human–animal welfare connection. Researchers and nurses are increasingly interested in the human–animal bond and how it can benefit human health. Public health agencies, like the American Public Health Association, are asking for a moratorium or ban on intensive animal agriculture in light of its enormous negative environmental impact.

There are doctors and scientific centers, like the Johns Hopkins Center for Alternatives to Animal Testing, which are finding and promoting superior non-animal testing methods. These first steps, as exciting as they are, only begin to shine the light on the path ahead and on the work yet to be done. In the final chapter, a way forward and a paradigm shift in how public health views the status of and our relationship with other animals is suggested. The future of public health is clear— to improve our health we must realize that the welfare of animals is intricately linked to our own.

2
Victims of Abuse: Making the Connection

> He who is cruel to animals becomes hard also in his dealings with men. We can judge the heart of a man by his treatment of animals.
>
> —Immanuel Kant

Humans are animals, too: The link between violence toward humans and animals

In *Agnes Grey*, a fictional autobiography of a governess, published in 1847, Anne Brontë examined the connection between the oppression of and cruelty against women and animals in Victorian society.[1] The connection between the abuse of animals and of vulnerable humans is not just the stuff of fiction, though. Sociologists have long recognized a link between the abuse of animals and the abuse of humans, particularly women and children, and that both reflect a larger social struggle between social power and inequality.[2] Additionally, almost all abusers select victims who are smaller and physically weaker than themselves.[3] Recognition of this connection has led to historical precedents, which fostered the development of some of our most powerful laws that help protect against violence toward women, children and animals. One of the more interesting precedents involved the use of an animal protection organization to help a child.

US diplomat and philanthropist Henry Bergh is credited with starting the humane welfare movement in North America, which began raising awareness about the often perilous plight of both animals and children and took action to protect them. In 1865 he founded the American Society for the Prevention of Cruelty to Animals (ASPCA) in New York, the first animal protection organization in the USA.[4] In 1874 the abuse of a young girl, Mary Ellen Wilson, was brought to the attention of Bergh. After learning of the girl's physical abuse and neglect, social worker Etta Wheeler asked the New York City Police Department to intervene. The police refused, however, citing lack of authority without proof of assault. Numerous charitable organizations were also approached but they also cited lack of authority to intervene. Finally, Wheeler approached Bergh, arguing that since the child was part of the animal kingdom she was entitled to protection under the ASPCA. Bergh initiated an investigation into Mary Ellen's situation, which ultimately led to her being placed into protective custody and to the conviction of her caretaker.

As a result of the public outcry over Mary Ellen's case, laws were amended in New York to allow for the establishment of child protection organizations. In 1875, the New York Society for the Prevention of Cruelty to Children was established with Henry Bergh as vice-president and one of the founding members; it was the first child protection agency in the world.[5] Since this time, many animal welfare organizations have incorporated the protection of both non-human animals and children into their missions, referring to themselves as humane societies.[6]

Mary Ellen's case illustrates the connection between two extremely vulnerable groups in society—children and animals. It makes sense that laws designed to protect one group could be used to protect the other since in both cases the abused are utterly powerless to protect themselves. Many of the earliest animal protectionists were also leaders in the anti-slavery movements.[7] For them, the connection between violence toward one vulnerable group and that of another was evident. But there is an even more fundamental connection here than that of the abuse of vulnerable groups—cruelty to animals is associated with general indifference toward the suffering of others, humans and non-humans alike.

Case studies

History is replete with serial and other killers, including some of the most notorious ones, who are believed to have directed their violent tendencies toward animals prior to directing them toward humans. Jeffrey Dahmer killed and impaled the heads of frogs, cats and dogs on sticks prior to killing and cannibalizing humans.[8] Albert DeSalvo, aka the 'Boston Strangler', killed 13 women as an adult. In his youth, he trapped dogs and cats in an orange crate and shot arrows at them.[9] One of DeSalvo's favorite activities was to place a starving cat in a crate with a puppy and watch the cat tear the puppy's eyes out.[10] Dennis Rader, known by his famous signature as the BTK (blind, torture, kill) killer, wrote in an account of his childhood that he had hanged a dog and a cat.[11] During the trial of convicted sniper Lee Boyd Malvo, a psychology professor testified that the teenager, who had killed ten people with a rifle, had 'pelted—and probably killed—numerous cats with marbles from a slingshot when he was about 14'.[12] Edward Kemper, who killed eight women, including his mother, revealed in his trial that he had a history of abusing cats and dogs as a child. For example, he buried a cat alive, dug the cat up after she had died, then displayed her head in his bedroom.[13] At the age of ten, he killed a cat with a machete and stored her dismembered parts in his closet. One of the US's most prolific killers, Carroll Cole, admitted that his first act of violence was strangling a puppy.[14] In 1985 he was executed for 5 of the 35 murders of which he was accused.[15]

In recent years we have seen a disturbing increase in deadly violence in our schools. In most cases, cruelty to animals preceded these violent outbursts. Friends of high school killer Kip Kinkel claimed that long before he walked into his high school cafeteria and opened fire on classmates, he boasted about killing cats, squirrels and chipmunks by putting lit firecrackers in their mouths.[16] In his diary, high school killer Luke Woodham described how, five months before embarking on his killing spree, he had killed his dog.[17] He wrote about how he and a friend tied his dog up in a plastic bag and beat her while she howled in pain. They then covered the bag with lighter fluid and set the dog on fire. Before one of the most notorious shootings in the USA, Columbine, Colorado, high school students Eric Harris and Dylan Klebold bragged to their friends about mutilating animals.[18] They later shot and killed 12 classmates and a teacher and injured more than 20 others before killing themselves.[19] 'There is a common theme to all of the shootings of recent years,' says Dr Harold S Koplewicz, Director of the Child Study Center at New York University. 'You have a child who has had symptoms of aggression toward his peers, an interest in fire, cruelty to animals, social isolation, and many warning signs that the school has ignored.'[20]

One of Australia's most brutal serial killers, Martin Bryant, murdered 35 people in a 19-hour killing spree in 1996. Later it was revealed that, as a child, he was referred to mental health agencies for torturing and harassing animals and tormenting his younger sister.[21] In the UK, one of the Moors Murderers, Ian Brady, tossed stray cats out of high apartment windows and watched them crash and die on the pavement.[22] He and accomplice Myra Hindley sexually assaulted and killed five children.

Examples of horrendous brutality toward both humans and non-humans are often cited by proponents and opponents alike of the animal–human interpersonal violence connection. Critics of the connection may argue that these are examples of only the most extreme cases and are, at best, anecdotes. Certainly, many methodological problems exist with these case studies, including lack of a comparison sample, reliance on second-hand sources and retrospective questioning. Although case studies lack the ability to establish causality, over the past 30 years scientists and criminologists have increasingly begun to research the animal–human violence connection.[23] As a result of these investigations, the relation between many forms of violent behaviors toward humans and violence toward animals is becoming indisputable. Violence and crime are recognized as public health issues in that they can affect the mental and physical health and the emotional security of both individuals and society at large. This chapter will present a few of the investigations illustrating a striking relationship between domestic violence, child abuse and other forms of criminal behavior and cruelty toward animals and why public health must also recognize this relationship.

Animal abuse and criminal behavior

Robert K Ressler, who developed profiles of serial killers for the Federal Bureau of Investigation (FBI), once stated, 'Murderers ... very often start out by killing and torturing animals as kids.'[24] Despite numerous case studies of murderers suggesting a link between cruelty toward humans and animals, there have been relatively few academic studies published on the topic.[25] Criminologists and sociologists have largely overlooked animal cruelty as a topic worthy of investigation, and it was not until 1997 that such work appeared in any criminological journal.[26] Despite the scarcity of studies, the ones that have been conducted suggest that childhood or prior history of animal cruelty is highly correlated with future, other violent and/or criminal acts. Animal cruelty can manifest as early as age six and is considered one of the most reliable predictors of later violent behavior.[27] Convicted criminals have frequently committed animal abuse, including drowning, shooting, strangling and smothering.[28] Animal abusers have killed cats by hanging and dogs and cats by exploding them or burning them alive. Other methods of animal abuse have included limb amputation, decapitation, brutal beatings, fracturing bones, stabbing and scalding with hot water. The most common animal victims include dogs, cats, rabbits and birds.[29]

A study of 261 incarcerated male inmates at medium- and maximum-security prisons found that 43 percent had engaged in animal cruelty.[30] Of these, 63 percent reported they had hurt or abused dogs and 55 percent had abused cats (some had abused both). The study researchers examined motives for the animal abuse and found that those who had abused animals out of anger or for fun, as opposed to out of dislike for the animals or imitation of others' behaviors, were more likely to repeat the violent crime for which they were convicted (rape, murder or assault).[31] Additionally, the analysis found that of all the animal abuse motivations analyzed, abusing an animal for fun in youth was the most statistically notable predictor for interpersonal violent behaviors as adults.

A comparison of a randomly selected sample of 45 violent offenders with 45 non-violent offenders at a maximum security penitentiary in Florida found a significantly greater incidence of history of cruelty to animals in the violent than in the non-violent offenders (56 versus 20 percent).[32] The investigators verified the information by using both interviews and institutional records. Another study divided 117 inmates into violent and non-violent offenders. The violent offenders engaged in animal cruelty significantly more often than the non-violent offenders (63 versus 11 percent).[33] Kellert and Felthous conducted interviews with 152 criminals and non-criminals in the states of Kansas and Connecticut.[34] They further divided the criminals into aggressive and non-aggressive, based on behavioral characteristics. Childhood abuse of animals occurred to a significantly greater degree among the aggressive criminals than either the non-aggressive criminals or the non-criminals.

In a survey of 64 inmates, 48 percent of those convicted of rape and 30 percent of those convicted for child sexual molestation had histories of cruelty to animals.[35] A case series by Johnson and Becker describes nine histories of adolescents with severe sadistic sexual fantasies. Of these nine, three had histories of violence against animals.[36] Ressler and other FBI investigators examined the behavioral traits and psychological motivations of convicted, incarcerated sexual murderers.[37] Of 28 individuals, 36 percent and 46 percent had committed animal cruelty in their childhood and adolescence respectively.

An investigation of incarcerated males revealed that those convicted of crimes against humans were more likely to have committed acts of animal sexual abuse in childhood or adolescence than other incarcerated males.[38] On the other hand, in a study of 137 rapists and 132 child sexual abusers, rapists were much more likely to have committed acts of non-sexual animal cruelty and child sexual abusers were much more likely to have engaged in sexual activities with animals.[39] In Australia, a study of convicted animal abusers found that more than 60 percent had also been convicted of violent crimes against people.[40] A police study in Canada found that 70 percent of those arrested for committing animal cruelty had histories of other violent crimes.[41] A similar study performed by the Chicago Police Department found that of 332 animal cruelty arrests, 86 percent of suspects had histories of multiple arrests, 70 percent had prior arrests on felony charges, 68 percent had prior arrests for drug sales or trafficking, 65 percent had been charged with battery, 59 percent were suspected gang members, 27 percent had firearm charges and 13 percent had arrests for sex-crimes.[42]

One of the most notable studies examining the connection between animal cruelty and other violent crimes was a collaboration between the Massachusetts Society for the Protection of Cruelty to Animals (MSPCA) and Northeastern University.[43] The investigators identified 153 convicted animal abusers prosecuted by the MSPCA between 1975 and 1996 and examined whether they had histories of other criminal behavior. The investigators found that in comparison with a control group, 70 percent of prosecuted animal abusers had also committed other crimes within the prior ten years, including interpersonal violent crimes, property damage and drug offenses. The three major findings of the study concluded that animal abusers are five times more likely to commit violence against people; four times more likely to commit property crimes; and three times more likely to be involved in drunken or disorderly offenses.

Animal cruelty is not just an act performed by individuals. Animal-fighting rings in which animals are trained to fight, often until death, occur worldwide and are conducted by gangs both covertly and overtly. The prosecution of US football player Michael Vick in 2007 brought to public light extremely cruel treatment of pit bull dogs, including horrific details involving abuse, torture and execution of underperforming dogs. Dogs were starved to make them more violent toward the opposing dog in a fight.[44] Fights ended when one dog died, or when a dog gave up. Vick executed dogs that did not perform well by hanging, drowning, strangulation, electrocution, shooting and/or slamming them on the ground.[45] A search of Vick's property uncovered the graves of seven of the dogs killed.[46]

Vick's fighting ring was hardly an isolated incident, however, and animal-fighting rings continue in parts of the USA and throughout the world. Dog fighting, cock fighting and other similar activities are connected to additional forms of crime, including money laundering and drug trafficking. 'Crime doesn't happen in a vacuum,' says John Goodwin of the Humane Society of the United States. 'When you have violent people betting large sums of money, you're going to have problems. Dog fighting is heavily linked to gambling, drugs, prostitution, gangs, and guns.'[47]

In addition to other criminal acts, animal cruelty is also associated with antisocial behaviors and personality traits, and non-criminal but destructive behaviours, such as substance abuse.[48] Using the results from a recent National Epidemiological Survey on Alcohol and Related Conditions (NESARC) in the USA, investigators examined the sociodemographic, psychiatric and behavioral correlates of cruelty to animals.[49] NESARC is a nationally representative sample of more than 43,000 non-institutionalized adults who were interviewed face to face. Embedded in the antisocial personality interview module was the following question about animal cruelty: 'In your entire life, did you ever hurt or be cruel to an animal or pet on purpose?' Respondents who answered yes to this question were defined as having a history of animal cruelty. The investigators found that cruelty to animals was significantly associated with all assessed antisocial behaviors, particularly robbery, harassment and forcing someone to have sex. In addition, pathological gambling, history of conduct disorder in childhood, obsessive–compulsive and histrionic personality disorders, and lifetime alcohol use disorders were strongly associated with a history of animal cruelty.

Another study by Gordon et al. looked at age of onset of substance abuse and its association with animal cruelty. They interviewed 193 adolescents entering outpatient substance abuse treatment centers.[50] Their study found a significant association between early onset of substance abuse and criminal involvement and cruelty toward both animals (defined as being physically cruel to animals two or more times in the previous year) and people. Cruelty to animals may also indicate mental illness, such as Munchausen's Syndrome, severe depression, dementia and schizophrenia.[51] Animal cruelty today is accepted as a diagnostic criterion for conduct disorder and anti-personality disorder in the *Diagnostic and Statistical Manual of Mental Disorders*.

Domestic violence, child abuse and the battered pet syndrome

One of the strongest associations between violence against animals and against humans is seen in domestic violence and in child abuse. A study of 860 college students suggests that animal abuse may be a 'red flag' for family violence.[52] Animal abuse, particularly abuse of companion animals, occurs disproportionately in households with inter-human violence.[53] In one study, 60 percent of the subjects who either witnessed or committed violent acts against animals as children reported histories of child maltreatment or domestic violence.[54] This includes child physical and sexual abuse, inter-sibling abuse and partner abuse (same sex or heterosexual). In a study in North Carolina in the USA, investigators compared police reports for disturbances and domestic violence and assault with animal cruelty reports.[55] They found that almost all the animal cruelty reports came from the same residences as the police reports.

The following is a typical case illustrating the linkage between child abuse and animal abuse. One morning, an Atlanta contractor pulled up to a house where he was to perform some work.[56] As he got out of his truck, he heard a dog screaming from the house next door, went over to investigate and saw through an open garage door a dog dragging his back legs and a woman standing beside him. The contractor intervened and took the dog to a veterinarian, whose suspicions about the incident were confirmed. The dog could not be saved and an autopsy revealed that the dog was paralyzed from having been beaten so badly. The incident was reported to the police. When the police went to the woman's house to make an arrest for abusing the dog, they found a badly bruised boy. Both parents were arrested for child abuse.

Companion animals are increasingly viewed as family members with inherent worth. More than 70 percent of US households with young children have companion animals.[57] In one study, seven- to ten-year-old children named on average two companion animals each when listing the ten most important individuals in their lives.[58] When asked 'Whom do you turn to when you are feeling sad, angry, happy or wanting to share a secret?', nearly half of the five-year-old children in another study mentioned their companion animals. Harm to companion animals can cause tremendous grief and anxiety in those who care for them. Unfortunately, their status as family member renders companion animals vulnerable to abuse, often as a means to exert control and intimidation over other humans. For example, an abusive father may hurt the family dog in order to scare his spouse or children into submission. Threats toward and actual abuse of animals in domestic violence situations occur for a variety of reasons, including:[59]

1. to confirm power and control over the family;
2. to perpetuate an environment of violence and fear;

3. to coerce the victim or prevent him or her from leaving;
4. to force the victim into silence;
5. to punish the victim; and
6. to further degrade the victim by forcing his or her participation in animal cruelty acts.

In a survey of 107 battered women, 47 percent of those with companion animals reported that their abusers threatened or harmed the animals.[60] Additionally, more than half of these women said their companion animals were important sources of emotional support and 40 percent had delayed seeking shelter out of concern for the animals' welfare. Once in the shelters, many of the women continued to worry about the animals' safety. That concern is not unfounded. Several cases reveal the horrific cruelty inflicted on animals by batterers: a pet cockatiel was beheaded because he was 'singing too much', a cat was hung by a leash, another cat was put into a microwave and other animals have been kicked, stabbed, shot or thrown.[61] In another study of battered women, 71 percent of those with animal companions reported that their partners had been violent to the animals.[62] The women reported that their partners abused animals for revenge or to psychologically control them. Quinlisk reported findings of a survey conducted as part of a domestic violence intervention program.[63] Of the 58 female victims of domestic violence who had companion animals, 68 percent reported violence directed toward their companion animals. In 88 percent of cases, the violence was committed in their presence. In 76 percent of these cases, their children also witnessed the animal cruelty. In other cases, women reported receiving threats either to kill or give away the animals.

The *New York Times* reported on a common scenario in which a violent partner abused a family companion animal to exact revenge against another.[64] In this incident, a man violently killed a dog who belonged to the female friend of a woman who had recently left him. The woman and her two children from a previous marriage were living with the friend. Since the female friend was housing the man's estranged partner, her dog became for him the optimum vehicle for expressing his rage against both women. 'He tortured the puppy when the two women weren't home,' said veterinarian Melinda Merck. 'He also tried to make two of the kids participate just to make it more heinous. So along with the animal cruelty, of course, we had child abuse.'

A study in Australia compared the experiences of 102 women recruited through various services intended to help victims of domestic violence with those of 102 women without histories of domestic violence.[65] The study produced three significant findings:

1. rates of partner or other family member companion animal abuse and rates of partner threats of abuse against companion animals were significantly higher in the families with domestic violence than those without;
2. as reported by their mothers, children from the violent families witnessed significantly more animal abuse than children from the non-violent families; and
3. women whose partners had threatened their companion animals were five times more likely to have experienced intimate partner violence.

A study comparing women residing at a domestic violence shelter with those without intimate partner violence revealed that women from the former group were almost 11 times as likely to report that their partner had harmed or killed their companion animals than women in the latter group.[66] In particular, severe physical inter-partner violence was found to be a significant predictor of companion animal abuse. Additionally, the majority of the shelter women and their children were emotionally close to the animals and were distraught by their abuse. In a survey of 48 of the largest shelters in the USA for victims of domestic violence and child abuse, more than 85 percent said that women who came in reported incidents of animal abuse and 63 percent of the shelters said that children who came in reported the same.[67]

Simmons and Lehmann examined how domestic violence consists of an array of controlling behaviors intended to intimidate and punish, including sexual, emotional, economic and physical abuse, isolation and threats.[68] In October 2007, a man killed the family dog by slitting the dog's throat after his wife asked for a divorce.[69] As this one case illustrates, harming animals is a common means of punishment against women and children who resist their abusers. In their survey of reports of 1283 women who had companion animals and were seeking refuge from male partner abuse, Simmons and Lehmann found that abusers who also harmed animals used more forms of violence and greater use of controlling behaviors than abusers who did not harm animals.

Four out of five battered women in a study in Wisconsin reported that their partners had harmed their companion animals, usually in their and their children's presence.[70] Additionally, many of the partners threatened to give the animals away as a means of exerting control over the women. A survey of college students found that those who had committed animal abuse as children had a significantly more favorable attitude toward corporal punishment of children and toward husbands slapping their wives.[71] No difference based on gender was found. The authors concluded that 'engaging in childhood violence against less powerful beings (animals) may generalize to the acceptance of violence against less powerful members of families and society—women and children'.

Escaping harm: The need to shelter all victims

Battered women often refuse to escape their abusive situations because they worry about the repercussions for their companion animals if they left. In fact, among women with companion animals, threats of harm against the animals represent one of the most common reasons for women to delay seeking shelter from partner violence. A Scottish Women's Aid spokeswoman said, 'it can be a case of a woman being told, "If you leave, I'll kill the cat or dog" and it is a very real threat. Some- times the children don't want to leave the pet.'[72] In one study, almost half of 41 battered women who had companion animals reported that their partners had threatened or harmed the animals and this threat was significantly associated with their decision to leave or stay with their partners.[73] In another, of the women whose partners had threatened or harmed their companion animals, 18 percent delayed seeking shelter out of fear for the animals' safety.[74]

The Scottish Society for the Prevention of Cruelty to Animals has han- dled cases in which animals have been beaten, hanged or even set on fire for revenge.[75] Further studies reveal that women frequently delayed their leaving abusive relationships out of concern for their companion animals.[76] Some even left their refuge and returned to their homes to check on the safety of their animals.[77] In one recent study in Australia, 102 women with a history of family violence were recruited through a domestic violence center and interviewed.[78] The study found that in comparison with a control group of women with no history of fam- ily violence, 53 percent of the women in violent relationships reported their companion animals had also been abused. By comparison, only 6 percent of the women in the control group reported harm to their companion animals, which was, in most cases, accidental. Addition- ally, 33 percent of the women in abusive relationships delayed leaving their partner by up to eight weeks out of concern for their animal companions. In 17 percent of these violent households the animals had been killed, and children witnessed animal abuse in 29 percent of cases.

Because most women's shelters do not allow animals, women are often left with no options for housing their companion animals until they establish their own homes.[79] Even if temporary housing could be provided for the women, they often refuse to be separated from their animals.[80] Many women would no more abandon their animals to harm than they would any other vulnerable family member. Additionally, many recognized that the protection and comfort their companion animals provided (whether emotionally or physically) often put the animals at greater risk of harm. For example, one victim of domestic violence stated, 'the dog knew when I was upset and would come stand in front of me to "guard" me, which enraged my husband all the more'.[81]

Animals often provide domestic violence victims with their only source of comfort and companionship during violent times.[82] In 2008, Dr Ann Fitzgerald of Windsor University in Ontario published a study titled 'They Gave Me a Reason to Live: The Protective Effects of Companion Animals on the Suicidality of Abused Women' in the journal *Humanity and Science*, which found that the presence of animals can both help women and heighten their risks of danger.[83] Fitzgerald noticed that 'for some abused women, their pets provide them with the support they need to cope with the abuse, which may result in their staying with their partner longer than they think they otherwise would have'. Fitzgerald concluded that women's shelters must start to accept companion animals.

'We had the experience with several women who would arrive with a garbage bag full of possessions and a pet in tow and refuse to check in when they learned that we would find a safe place for the animal, but it couldn't stay here with them,' one advocate for battered women said.[84] 'We've known of women who lived in their cars so they could keep their pets with them.' Prompted by numerous similar cases, there is now a movement to create women's shelters that provide on-care sites for companion animals so that families, including the animal members, can stay together.[85]

USA Today reported one such safe haven.[86] The story described one battered woman whose ability to escape her abusive partner and pick up the pieces of her life was greatly facilitated by the ability to keep her companion animal with her.

When Rose Terry finally resolved to leave her abusive boyfriend, she knew she'd have to live in a shelter for a few weeks before she could start life anew. She had no reservations about that. But she anguished over Byron, the cat who had seen her through the awful times. None of her friends could take the female feline (the family was first told she was male, hence the name), and she couldn't bear the thought of placing her in an animal shelter until she got back on her feet. 'I was desperate, weeping,' Terry says. 'She's my family.'

When Terry learned one Las Vegas domestic violence safe haven, Shade Tree Shelter, had just built a pet-boarding facility on its grounds for residents' animals, 'I was in such relief.' Terry packed up her suitcase and her cat just before Christmas and checked in. 'It's just so good to get to visit with Byron every day,' says Terry, 55, who has a new job and nearly enough savings to lease an apartment and start over. 'It helped so much that I didn't have to worry about her.'

The American Humane Association has created the Pets and Women's Shelters Program start-up guides to assist domestic shelters in providing animal housing.[87] In the UK, Paws For Kids helps foster companion animals for women seeking refuge.[88] However, the numbers of shelters providing such services are still only a handful worldwide, and there is great need for more.[89] Battered women have been known to live in their cars with their animal companions for as long as four months until an opening became available at a pet-friendly safe house.[90] Many battered partners have stated that they would have left their abusive relationship much earlier if there were facilities to accommodate their companion animals.[91] Children also benefit from such facilities. As one women's advocate explains, 'pets are very important to women and children experiencing domestic violence and they can have very real difficulty when they can't take them'.[92]

Cycle of violence: Children and animal abuse

Providing incentives for women and children to leave violent households earlier by offering combined human-animal shelters not only helps all victims (human and animal) but may also help break the cycle of violence. Childhood abuse of animals is considered one of the most reliable indicators of family violence and violence toward children. Children who abuse animals are most often from dysfunctional and/or violent families, where they are frequently witnesses to, or victims of, domestic violence. Such children are often subjected to parental neglect and/or physical, mental or sexual abuse.[93] Felthous investigated animal cruelty incidence in 346 male psychiatric patients who were categorized according to their level of aggression.[94] He found that those who had been cruel to animals had suffered from extreme parental punishment and destructive behavior. A survey of 267 college undergraduate students found that males with histories of physical paternal punishment more frequently committed animal cruelty than males who had not experienced such punishment.[95] This association did not hold for female students or for maternal punishment.

A community sample in Canada of 47 mothers with a history of domestic violence was compared with matched controls.[96] Children from families with a history of domestic violence were significantly more likely to have been cruel to animals than children from non-violent families. In another study, of 164 battered women matched with 199 control women, children from violent homes were 2.3 times as likely to be cruel to animals as children from non-violent homes.[97] As these studies suggest, preventing childhood exposure to domestic violence may reduce their odds of becoming perpetrators of violence themselves.

Several studies suggest that children who are themselves the victims of abuse may be more likely to harm animals. A study including more than 1000 sexually abused and control children found that 18 percent of the sexually abused children were cruel to animals compared with 3 percent of the control children.[98] Maternal caregivers of 1433 six- to twelve-year-old children completed questionnaires on their children's behaviors and provided information about domestic and sexual abuse. The children were then divided into three groups: a normative sample, a sexually abused sample and a psychiatric comparison group without sexual abuse histories. Violence against animals was significantly associated with violence against humans in all three groups, and the prevalence of animal violence was more than five times as high for the sexual abuse and psychiatric groups as for the normative group.

In one study, a shocking 88 percent of families in which children were physically abused also had histories of animal cruelty.[99] The fathers committed two-thirds of the animal cruelties and the children themselves committed almost one-third. Of 23 families with a history of violence against animals, 35 percent also involved children who were believed to be at risk of harm and were placed on the social services risk register.[100] A study was conducted in a residential treatment facility comparing the family histories of 50 boys with behavioral problems and who were cruel to animals with 50 boys with behavioral problems but were not cruel to animals.[101] The study revealed that boys who were cruel to animals had significantly greater histories of being sexually abused, being physically abused and being exposed to domestic violence than the boys who were not cruel to animals. A study of 286 college students revealed that 37 percent of males had engaged in animal cruelty and most (63 percent) initiated the abuse prior to the age of 13.[102] Additionally, men who reported a history of sexual abuse were more likely to report committing animal cruelty than those without sexual abuse histories.

Children may abuse animals for a variety of reasons, including identification with and imitation of the abuser, post-traumatic play and even killing an animal to protect the animal from further harm from the adult abuser.[103] Children who grow up in homes where companion animals are abused or neglected may be more likely to see such treatment as acceptable and emulate patterns of abuse demonstrated by their caregivers.[104] In a study reported by Quinlisk, more than half of the children who had witnessed animal cruelty emulated this behavior.[105] A number of additional studies have demonstrated that a violent home environment may be associated with childhood animal cruelty.[106]

As with domestic violence, animal cruelty may be used as a means to control sexual assault victims. For example, a teenager testified in court that her father threatened harm against her companion animal if she refused his sexual advances.[107] Other studies have revealed that child sexual abusers may threaten or harm companion animals as a means of silencing the child victims.[108] Abusers kill, harm or threaten children's companion animals to coerce them into sexual acts or to force them to remain silent about abuse. As a result of such intimidation, abused children may learn that abuse of others is a powerful means of exerting control. Disturbed children may kill or harm animals to emulate their parents' conduct, to prevent the abuser from killing the animal or to vent their own aggressions and frustrations on another victim.[109]

Just as some children from violent homes or who are direct victims of violence commit cruelty against animals, others suffer emotional distress witnessing cruelty toward animals. Children from violent homes generally follow two patterns: they either begin to abuse the animals themselves or they bond strongly with their companion animals.[110] Often, children from violent homes rely heavily on their family animals for companionship and comfort.[111] Unfortunately, animals from such homes are often killed, die from neglect or run away and seldom live beyond the first few years of life. As a result, there may be constant turnover of companion animals and the children may suffer from repeated cycles of attachment and loss.[112]

It is unclear why experiencing or witnessing animal abuse may lead to greater attachment to animals in some children and cruelty to them in others. One theory is that witnessing infrequent or milder forms of animal abuse may lead to greater empathy, while frequent exposure of animal abuse desensitizes an individual to suffering.[113] Additionally, evidence suggests that the younger the children are when they witness animal abuse, the more likely they are to abuse animals themselves.[114]

School bullying

Adolescents, particularly males, may commit animal cruelty in response to social isolation or rejection, or to impress their peers.[115] Although few studies have explored this issue, the connection between bullying and animal cruelty may be quite strong.[116] In the USA, of the nine school shootings that took place between 1996 and 1999, half of the shooters had histories of violence against animals.[117] In addition to a history of animal cruelty, another common factor among them was a history of persecution, rejection, social isolation and other forms of direct or indirect bullying by school peers. In Italy, 268 girls and 264 boys aged 9–12 completed questionnaires about animal abuse, bullying and victimization at home and school.[118] Some 46 percent of boys and 36 percent of girls reported that they had been cruel to animals at least once. One-third experienced domestic violence, and over one-third had been abused by one or both parents. Additionally, two in five children had been victimized at school, either directly (i.e. physically hurt, threatened or called names) or indirectly (i.e. being isolated or having harmful rumors spread). An association between bullying victimization and cruelty against animals was found, mainly in boys. The authors concluded that 'discovery of animal abuse should prompt further enquiries about other problems that a child may have'.

Another study of 241 adolescents in Australia revealed that just witnessing animal abuse predicted bullying behavior.[119] A survey of 185 college males in the USA revealed that 30 percent had committed cruelty to animals at least once and most of them multiple times.[120] One-time animal abusers were not more likely to be victims of or perpetrators of physical or verbal school bullying, compared with non-animal abusers. However, those who committed at least two acts of animal abuse were more likely to be either victims or perpetrators of peer bullying. Those who were victimized or who bullied others the most frequently also reported the highest rates of multiple acts of animal cruelty.

Does violence toward animals lead to violence toward humans?

In most of the studies explored in this chapter, there are examples of many violent adults who have no known history of having abused animals. Thus, not all who have committed or desired to commit violent acts against people have committed violence against animals, and not all who conduct animal cruelty commit violent acts against people. Animal abuse may occur in isolation. Children often unintentionally or intentionally harm or kill animals as part of experimentation as they grow up, without ever committing violence against humans. Teens, particularly teenage boys, may harm an animal to impress peers, without committing further violence.[121]

In cases where a perpetrator committed acts of violence against both humans and animals, however, the question remains: Was there a developmental sequence? In other words, was violence toward animals a necessary prerequisite for later violence against humans? Do some people who harm humans learn first by harming animals? Theories abound as to why, if true, animal abuse may lead to later abuse against humans. Because animals have no voice or way to retaliate and the legal repercussions against harming animals is often minimal, conducting violence against animals is, on the whole, far easier for most than conducting violence against humans. This relative ease of harming animals may prepare or condition an individual to later cause harm to another human, often referred to as the violence graduation theory.[122] A second theory, dubbed the 'social learning theory', contends that children may learn, by witnessing or harming animals, that violence is an accepted means of expression.[123] There is much evidence to suggest that child-rearing environments influence engagement in animal cruelty. As discussed earlier, children growing up in violent homes often learn to imitate violence by abusing animals weaker than themselves and, through reinforcement, they become abusive to other humans.[124] Experiencing or engaging in animal cruelty may interfere with a child's development of empathy.[125]

The deviance generalization theory suggests that animal abuse is one of many manifestations of antisocial behavior that can develop from childhood onwards, which have the same underlying causes and which occur in no particular time order.[126] A review of the criminal records of 153 animal abusers convicted in Massachusetts from 1975 to 1996, and 153 controls, revealed that while animal abusers were more likely to be violent offenders than non-animal abusers, the animal abuse was no more likely to precede than follow the violent offenses against humans.[127] Nevertheless, among those who commit the most heinous abuses, there is stronger evidence that behaviors learned in childhood may prepare them for future violence against humans. A recent study examined the case files of 44 of the most sadistic serial killers—those who had tortured their victims prior to killing them.[128] Of these, almost two-thirds were reported to have previously injured or killed animals and half had tortured the animals before killing them. These killers took pleasure in causing their victims to suffer and often used the same techniques of torture against both animal and human victims. The common modus operandi used against both human and animal victims suggests that many serial killers learn their malicious behaviors as children with animals, then later apply what they have learned to their human victims.

Alternatively to the theories suggested above, strain and distress, such as what follows by being a victim of violence oneself, may motivate one to commit acts of cruelty toward animals or humans for the purpose of seeking revenge, reducing stress or managing negative emotions.[129] Animals may be an easy target for the discharge of aggression.

Limitations of current studies

Disentangling the cause and effect of violence is extremely difficult, particularly because of the nature of the studies that have been conducted. A meta-analysis was conducted on 15 controlled studies investigating the link between childhood animal cruelty and later violence against people.[130] Of these 15 studies, 5 supported the link but the remaining 10 did not. The investigators examined the possible methodological reasons for this discrepancy and found that studies that supported the link used direct interviews to examine subjects, used repeated acts of animal cruelty rather than one act as the dependent variable and explicitly defined animal cruelty and aggression toward people. The meta-analysis found that studies that did not support the link mainly relied on chart reviews for investigating subjects' experiences with animal abuse, used one act of animal abuse as the dependent variable and did not explicitly define the behaviors. Other factors that might affect the relationship between animal abuse and human abuse and that have not been consistently examined include age at onset, methods of animal abuse, motivation for abuse, frequency of animal abuse and type of animal abused.[131]

Measuring and assessing animal cruelty has been problematic for a variety of additional reasons. One of the most significant is that the definition of animal cruelty varies by investigator and has evolved over time, not unlike the definition of child abuse. One of the most frequently used definitions was developed by psychologist Frank Ascione, a leading authority on the connection between animal abuse, child abuse and domestic violence. As defined by Ascione animal abuse is a socially unacceptable behavior that intentionally causes unnecessary pain, suffering or distress to, and/or death of, an animal.[132] However, this definition has been criticized as being too narrow.[133] It omits, for example, behaviors that are socially or culturally acceptable but nevertheless cause suffering in animals.

Additional problems in defining animal cruelty include differing attitudes toward, and acceptance of, behaviors relating to different species, the range of severity from teasing to torture, passive (i.e. neglect, deprivation) versus active abuse, mental/psychological versus physical abuse, and whether or not the frequency of abuse should matter. As a result, definitions of animal cruelty used by investigators have ranged from that developed by Ascione, to those which include sexual abuse and neglect and passive cruelty exemplified by animal hoarders, many of whom believe they are acting out of good intentions.[134] Thus, as with the definition of child abuse, the definition of animal abuse has varied by culture, societal moral paradigms and time.[135]

In addition to the discrepancies in how animal abuse is defined, many reports of animal abuse are derived from second-hand sources. In particular, parents are often relied upon to report animal abuse by their children. Given that abuse of animals is often covertly committed, parents are often not aware of their child's behaviors.[136] Studies have indicated that parents significantly underestimate their children's involvement in cruelty toward animals. For example, in one study, parents' reports of their children's participation in animal cruelty suggested a prevalence of only 2 percent compared with the children's self-report suggesting a prevalence of 10 percent.[137] Anecdotal evidence also suggests that parents do not always take seriously their child's cruelty toward animals.[138] Relying on second-hand sources to investigate animal abuse is thus problematic. On the other hand, self-reporting of animal abuse is also problematic, being affected by the willingness of the perpetrator to admit the act.[139]

Additional complications in assessing animal cruelty include differences in the sensitivity of the survey instrument, underreporting and the scarcity of data on animal abuse in official records.[140] There are no large-scale, police-based data available on the incidence of animal cruelty in any nation.[141] Measurements to assess animal cruelty in childhood are limited and are most often included in measures designed to assess more general behaviors.[142] Furthermore, most studies that have been conducted included clinical samples and clinical instruments, which tend to measure physical cruelty. As a result, the types of animal cruelty captured by these studies tend to be among the most severe and likely underestimate other forms of animal cruelty, such as emotional abuse.

To truly answer the question of whether animal cruelty leads to cruelty against humans, much more information concerning the nature of the abuses needs to be ascertained. Despite more than five decades of research that has demonstrated that abuse of animals is an important indicator of mental illness and possibly a contributor to adult violence against humans, only a handful of studies have investigated the specific methods of animal cruelty and its prevalence. However, these studies suggest that cruelty to animals is quite common and the prevalence in the general population may be as high as 20–51 percent.[143] If this statistic is true of the general population, then it is reasonable to postulate that the prevalence may be even higher among those with certain psychological disorders and those with histories of violence toward other humans.

The UK's National Society for the Protection of Cruelty to Children (NSPCC) conducted a careful review of the existing research on animal abuse and the abuse of women and children.[144] It found six major emerging themes and summarized the findings of each:

1. Animal abuse perpetrated by children: 'Aggressive acts against animals can be an early diagnostic indicator of future psychopathology, which, if unrecognized and untreated, may escalate in range and severity against other victims.'
2. Acts of animal abuse witnessed by children: 'Exposure to animal abuse desensitizes children to violence. This desensitization may come through individual traumatic acts against animal companions, or through cultural conditioning.'
3. Acts of animal abuse in the context of domestic violence: 'Animals and children living in violent households may become victims of abuse themselves. Acts of animal abuse may be used in order to coerce, control, and intimidate battered women and their children to remain in, or be silent about, abusive situations.'
4. Animal abuse as part of the continuum of family violence: 'Animal abuse should not be regarded as an isolated incident with only an

animal victim but rather as an unrecognized component of family violence.'

5. Therapeutic potential of animals to promote healing or enhance empathy skills: 'Abuse victims may find interactions with a family pet a source of comfort.'

6. The role of animals in child development: 'Animal companionship can help children move along the developmental continuum.'

Animal abuse is a public health matter

Regardless of the causal sequence of animal and human abuse, it is clear that there is a correlation between the two that warrants action. The studies conducted indicate a common pattern of abuse against the most vulnerable, particularly women, children and animals. Although the study of convicted animal abusers in Massachusetts did not find evidence to support the theory that animal abuse leads to human abuse, it did provide compelling evidence that the two are interconnected.[145] The authors concluded that 'although these findings dispute the assumption that animal abuse inevitably leads to violence toward humans, they point to an association between animal abuse and a host of antisocial behaviors including violence'. The NSPCC review made a similar conclusion about animal abuse and child abuse, stating, 'Violence against animals cannot be dismissed or treated as an isolated problem. Rather, acts of animal abuse should be considered within the context of a much wider picture of family violence. Consequently policies, service provision, and training should take account of the link.'[146]

Despite the methodological flaws in some studies, the connection between animal abuse by children and other concurrent or future violent acts is also widely accepted. There is a general understanding that childhood animal cruelty should signal that a child may need clinical attention.[147] Indeed, the National Research Council and the FBI in the USA now recognize that childhood animal cruelty is a 'powerful indicator of violence elsewhere in a perpetrator's life'.[148] Although it is not connected to all cases of violence toward people, animal abuse should be regarded as a 'red flag' for other forms of abuse and violence. Detecting animal abuse early may help prevent other forms of abuse. Furthermore, protecting and securing the safety of animals may provide comfort for their human caretakers and help battered women, who might otherwise hesitate to leave their abuser out of concern for their animal companions, escape their abuse.

Generally, medical and public health professionals and agencies have not been very active in combatting animal cruelty. When the US Supreme Court recently reviewed a federal law aimed at banning video depictions of violence toward animals, there was a noticeable lack of response or involvement from the public health community. The federal law banned the sale of 'crush videos', which depict animals being crushed to death by women wearing stiletto heels or with their bare feet.[149] Public health action in support of this law may have prevented its overturn by the Supreme Court (done in the name of free speech) in 2010. Videos depicting child sexual abuse are illegal in the USA because the harms of such tapes override free speech concerns. Public health advocates could easily have used the same argument against crush videos.

Often, public health practitioners will deny the link between animal abuse and human abuse, or object to working on animal cruelty issues. Some human welfare advocates, for example, have expressed a need to keep their work for humans separate from any concerns about animals for fear of losing their focus.[150] Others feel that their strategy is weakened if they include concern for animals in their work or simply place animal cruelty issues as low priority. Yet, as one health advocate argued, 'practitioners should ... be more aware of the broader meaning and role of pets in family life because this can be an important dimension in understanding the patterns of relationships and beliefs in families'.[151] Dr Sherley of the School of Medicine, Australian National University, argued:

> There are good reasons why medical practitioners should be particularly concerned by animal cruelty. Both intentional and unintentional acts of cruelty may reflect underlying mental health problems. Cruelty within the family setting is an important sentinel for domestic violence and should prompt an assessment for possible child abuse.[152]

How we can help

Although there is substantial evidence connecting violence toward humans and animals, suggesting that urgent action should be taken, further research can help shed light on several factors: on the best methods to detect abuse, the psychological motivation behind violent tendencies toward animals and humans, and the most effective approaches to combat violent tendencies. One place to start is by understanding the true prevalence of animal cruelty, as it is largely unknown and likely to be vastly underestimated.[153] Larger and more systematic studies using controlled samples should be conducted since studies exploring the connection between violence against animals and against humans have mostly involved small sample sizes in the UK and the USA.[154] Most studies review animal cruelty at the individual level, ignoring social and cultural factors.[155] These limitations and gaps in knowledge suggest for us the types of studies that should be conducted next. Researchers must strive to publish larger systematic studies in public health journals. Most studies addressing violence toward animals are published in journals specifically geared toward the study of animals, such as veterinary journals. These have yet to cross over into mainstream, human-oriented publications, such as clinical psychology journals.[156] I found only a handful of articles on this issue in medical journals and none in public health journals. Additionally, funding for such studies is rare and often comes from humane organizations.[157] Health agencies can help fund such studies. Perhaps most significant of all, public health must recognize the animal–human abuse connection. Most of the largest health agencies, such as the WHO and US Centers for Disease Control and Prevention, do not make any reference to the connection between violence against animals and that against women and children in their reports on domestic violence and child abuse. This needs to change.

Animal cruelty should become a more integrated part of health education. Currently, with the exception of a brief mention of animal cruelty as a sign of antisocial personality disorder or conduct disorder, animal cruelty is not routinely discussed in medical or public health training. Practitioners should be trained to recognize potential animal cruelty being witnessed or inflicted by their patients, and be instructed in ways to notify the proper authorities and coordinate with social welfare institutions.

Greater coordination between human health, animal protection, veterinary and social service agencies will benefit all victims of violence. Historically, many animal welfare and child protection organizations began conjointly working toward a united goal to protect both groups of vulnerables.[158] However, institutional changes over the past century have led to the separation of child welfare from animal welfare organizations. Domestic protection agencies are also, as a rule, separated from animal protection services. Most child welfare agencies do not assess animal cruelty.[159] There is thus little cross-reporting between the agencies created to protect humans and those created to protect animals. As a result, a vital opportunity to prevent, detect and combat violence is lost.[160] Collaboration between the various organizations can greatly expand services and resources, as was determined jointly by the NSPCC and the Royal Society for the Prevention of Cruelty to Animals in the UK.[161]

Recognizing the connection between all forms of domestic violence, toward humans and non-humans, the NSPCC is taking a proactive approach in coordinating efforts with animal humane organizations, researching the connection and educating the public.[162] In Ontario, Canada, uniting a child welfare organization with an animal welfare organization improved the detection of violence against both children and animals.[163] In the USA, several states have enacted laws that require and authorize cross-reporting between child, spousal and animal abuse investigators.[164] Public health practitioners and officials can support such endeavors. And, as discussed earlier, the creation of safe havens for animal and human victims of violence is urgently needed.

To help prevent both animal and human cruelty, public health can promote and support programs designed to cultivate empathy for animals. A lack of empathy is closely linked to the abuse of animals.[165] A well-developed sense of empathy, on the other hand, corresponds with healthy emotional development in children. Research demonstrates a link between empathy directed toward humans and animals in both children and adults. In one study, empathy for companion animals in preschool children was correlated with empathy toward other children.[166] Children with the strongest bonds with animals had the highest scores for empathy for other children. Humane education is just in its early stages and is starting to become widespread.[167] These are typically school-based programs intended to teach children kindness toward animals with the hopes that such programs may foster respect and compassion for all living beings—human and non-human. Such programs have, thus far, yielded promising findings.[168]

Public health can urge the enactment and enforcement of stronger animal cruelty laws, including laws against animal-fighting and other types of abuses inflicted by groups of individuals. Crimes against animals are often ignored, given low priority, left unpunished or under-punished.[169] For example, in the USA, of 80,000 cases of animal cruelty filed between 1975 and 1996, only 268 were prosecuted.[170] Of these, most of the perpetrators simply paid a small fine. The crush video Supreme Court case was a ripe and missed opportunity for public health specialists to educate the public and policy makers on the connection between animal and human abuse. But there are other opportunities in the pipeline. California may soon have in place an online registry whereby adult animal abusers will be listed along with their home addresses and places of employment, similar to registries of sex offenders.[171] The intention is to be able to keep track of those who have been violent to animals to help prevent further violence against both humans and animals. Tennessee is considering a similar registry.[172] In March 2006, Maine Governor John Baldacci signed a law, the first of its kind in the USA, which permits judges to include animal companions in court-issued protection orders against domestic abusers.[173] Other states, including Vermont, New York, California and Colorado, followed suit.[174] With the support of health specialists, similar regulations could be implemented worldwide.

Lastly, regardless of animals' usefulness in identifying other forms of abuse and facilitating the escape from abuse, their abuse is connected with larger issues about empathy and violence. As Dr Sherley commented,

> It should not just be a desire to minimize aggression toward ourselves that motivates us to oppose animal cruelty. Acts of cruelty are inherently wrong; they lessen us as a society. Animals, children, the aged, the ill, the disabled, and the marginalized are all subject to victimization and are deserving of society's protection.[175]

Violence against animals raises important questions about the nature of empathy and kindness and the type of society that we wish to live in. As advocates for a better, healthier society, it behooves us to protect all, particularly the most vulnerable, who are victims of abuse.

3
Lions, Tigers and Bears:
The Global Trade in Animals

> When you have got an elephant by the hind legs and he is
> trying to run away, it's best to let him run.
>
> —Abraham Lincoln

An elephant never forgets

On 20 August 1994, an international uproar was caused by an incident involving a female African elephant who went on a rampage in
Honolulu, Hawaii.[1] The animal, Tyke, was 'performing' during an event
for Circus International when, before hundreds of horrified spectators,
she grabbed her trainer, thrashed him about and killed him before turning on her groomer and goring him. She then ran from the arena and
escaped to city streets, where for 30 minutes she caused havoc and
threatened the public before police shot her almost 100 times. It took
her two hours to die.[2]

Why did Tyke, after years of performing for this circus, suddenly turn
and attack the two people with whom she had spent most of her circus
life? To answer this, it might help to take a look at the life she led up
until that fatal day. Tyke's keeper, John F. Cuneo Jr., owned Hawthorn
Corporation, one of the largest suppliers of performing elephants and
tigers in the USA.[3] While most of Tyke's life history with Hawthorn
Corp. is not publicly known, that of Lota, another elephant kept by the
company, is. In 1952, she was captured from the wild as a baby in India
and torn from her family.[4] Lota lived her first two years in captivity in
a zoo in India before being shipped to the Milwaukee County Zoo in
the USA, where she spent the next 36 years of her life with three other
female elephants.[5] At the Milwaukee zoo, the three elephants were routinely chained by two legs to the floor of their barn for at least 18 hours a

52

day. Zoo staff conducted videotaped training sessions for new employees in which the elephants were repeatedly struck by bullhooks.

Over time, Lota became too aggressive for the zoo to handle. In 1990, she was sold for US$1 to Hawthorn Corp.[6] In a widely publicized video, she was shown being beaten and dragged into a trailer as she fought her chains, which finally broke, sending her falling backwards and then sliding beneath the trailer. This video footage caused an international outcry and repeated pleas that the elephant be sent to a sanctuary. Despite the pleas, Hawthorn would not relinquish Lota and she was kept in chains throughout her life, being dragged around and rented out to one venue after another to perform. In 1996, she contracted tuberculosis. In 2001, a US Department of Agriculture (USDA) inspector cited Hawthorn for failure to provide veterinary care to Lota, who was 'excessively thin, with a protruding spine and hip bones and sunken in eyes'.[7] No improvement in her condition was made, however. Returning later that year, the USDA again noted Lota's dismal state, reporting that she was in a 'perilously emaciated state, with a wound on her left hip'. The elephant died from tuberculosis in 2005.

Based on Lota's experiences, it seems likely that Tyke's life was similarly wretched under the care of Hawthorn Corp. A report in *Nature* reveals that elephants, when exposed to violence and psychological and social trauma, can suffer from post-traumatic stress disorder.[8] Could it be that Tyke, after a lifetime of physical and mental suffering, intentionally lashed out and sought revenge upon those who had harmed her? Or was her attack the result of general psychological illness produced by years of confinement, captivity in sterile environments and physical abuse? We will never know why Tyke lashed out, but we do know that Lota's experiences are by no means an exception. Today, thousands of wild or exotic animals are kept in zoos, circuses, marine amusement parks and private residences. Many are used to supply hunting ranches and game parks around the world. These animals are either caught from the wild or bred in captivity and traded around the globe to ensure an ever-ready supply. This is the global trade in wildlife, and all indications are that our infatuation with exotic and wild animals is coming back to bite us.

The wildlife trade

Unlike dogs, cats and cows, who have been domesticated over centuries, a wild animal is one who has not been domesticated to live with humans. Thus wild animals include not only free animals who are then captured but also animals who have, only in relatively recent times, been bred in captivity by humans. In this chapter, we will explore not only how the global wildlife trade poses significant public health risks but also how it causes immense suffering in animals. The trade is directly and indirectly leading to a rapid rise in new infectious diseases, the spread of existing diseases, injuries in people and the loss of species at a rate never before seen. As we delve deeper into the world's forests and jungles to capture, kill and collect animals for the trade, we are inviting pathogens never before encountered to jump into the human population and wreak havoc. As we rip animals from their natural habitats, we are disrupting ecosystems in profound and perhaps irreversible ways, which will in turn cause a surge in some very deadly infectious diseases. And, as we ship billions of animals around the globe, we are ensuring that any diseases unleashed by this trade will impact humans everywhere.

Every year, billions of animals are caught from the wild or bred in captivity and then traded or slaughtered in the wildlife trade.[9] They are used live or sold in body parts as pets, entertainment, food, skins, ornamental or medicinal objects, and biomedical research subjects. They are also stocked on hunting ranches to be slaughtered in 'canned' hunts, in which hunters pay fees to shoot and kill exotic animals in a confined area from which the animals are unable to escape. The global trade in animals involves an unprecedented number and array of species, including non-human primates (NHPs), other mammals, birds, amphibians and reptiles. The trade occurs both globally and within countries, and all indications are that it is rapidly increasing worldwide.[10] The drive behind the exotic animal trade is big money, estimated at anywhere from US$10 billion to more than US$40 billion annually.[11]

The USA is by far the largest consumer of wildlife.[12] An analysis of the US Fish and Wildlife records revealed that the USA imported more than a billion animals, including fish, between 2000 and 2004.[13] Excluding fish, more than 180 million animals were imported during that time. Another analysis reports that among mammals, about two-thirds are imported for commercial purposes and one-third for biomedical research.[14] In addition to being the largest importer of wildlife, the USA is also, paradoxically, the largest exporter of wildlife.[15] For example, between 1989 and 2007, the USA exported almost 58 million live reptiles.[16] The number of exports of reptiles from the USA doubled between 1989 and 2007. After the USA, the world's other wealthiest and most developed nations, including those in Europe, China and Japan, are not far behind in their consumption of wild animals.[17] A review of customs data in China revealed a sharp increase in the number of live turtles imported between 1998 and 2002;[18] in 2002 more than 2 million live turtles and tortoises were imported. Other nations have similarly been seeing an increase in the importation, exportation and/or regional trading of wildlife over recent years, particularly in Southeast Asia, a 'hot spot' of exotic species.[19]

The Convention on International Trade in Endangered Species (CITES) has regulated the trade of wildlife across borders since 1973.[20] Its purpose is to prevent the trade from threatening endangered animals and plants.[21] The treaty contains three appendices in which restricted species of plants and animals are listed. The degree of restriction placed on the trading of a species depends on which appendix that species is listed in. In all, about 5000 animal species are listed in any of the three appendices. Although nations may supplement CITES agreements with further regulations or bans on importation and exportation of animals, any enforcement of CITES relies largely on individual nations, many of which lack sufficient resources to uphold restrictions.[22] Thus, in reality, only a limited number of regulations protecting the listed species are actually enforced.

Simply put, virtually any animal is fair game and no region in the world is immune from the health repercussions of the trade. This chapter will focus on three main issues concerning the wildlife trade:

1. how wild animals kept as pets and entertainment are causing injuries and death, particularly in children;
2. how the wildlife trade is contributing to the emergence of some of our deadliest infectious diseases; and
3. how the trade causes the loss of biodiversity that, in turn, has rippling public health effects.

Our fatal attraction

Very often, the more exotic or wild the animals, the more intriguing they seem to us. It's hard not to be charmed by a tiger or even a baby iguana. Unfortunately, their allure can lead to fatal consequences. There have been numerous reports about horrifying attacks by wild or exotic animals kept as pets, housed on hunting ranches or used for entertainment. There are countless stories around the world like the following:

• In December 2003, a bear who performed at a children's theater in Moscow killed a trainer as he entered the animal's cage to feed him.[23]

- In 2008, a grizzly bear named Rocky, who appeared in the Will Ferrell movie *Semi-Pro*, was being filmed for a promotional video for the wild animal training center. The bear attacked and killed his handler by taking a lethal bite out of the man's neck.[24] Prior to this, Rocky was described as a 'loving, affectionate, friendly, safe bear'.
- In January 2010, a Canadian man who kept exotic cats in cages behind his farmhouse went out to feed a Bengal tiger and never returned.[25] Several years earlier, he had challenged a ban on keeping exotic animals after one of the tigers kept on his premises attacked a ten-year old boy during a photo opportunity.[26] The man won the court case and had the ban overturned, but he lost his life in 2010: he was found dead in the tiger's cage.

These examples are just a tiny sample and, unfortunately, children are especially vulnerable to such attacks. Because of their smaller size, kids are much more likely to be seen as prey. For example:

- In April 2009, a wolf performing on stage in California for a show by the company Amazing Animal Productions lunged at a two-year old girl in the audience, biting her neck and face.[27] Fortunately she survived, but other children have not been so lucky.
- In August 1999 in Illinois, a three-year-old toddler was strangled to death by the family's 7.5 foot African rock python.[28]
- In 2003, a ten-year-old boy was shoveling snow near a cage containing a 400 lb Bengal tiger outside his aunt's home.[29] The boy got too close and the tiger dragged him under a fence, into his cage and mauled him to death.

Despite numerous similar stories of severe human injuries and death, the trade in wild animals as pets and entertainment continues to flourish and expand. The global wildlife trade of pets alone is estimated to involve at least 350 million live animals annually.[30] Here again, the USA takes the lead in the number of wild animals kept as pets. According to a survey conducted by the American Pet Product Manufacturers Association, 18.2 million wild animals were kept as pets in 2004, an increase from 16.8 million in 2002.[31] Thousands of wild mammals, such as tigers, lions, wolves and NHPs, are kept in American homes.[32] Born Free USA, a non-profit animal protection organization, estimates that between 5000 and 7000 tigers are kept in private homes in the USA.[33] If correct, this means more tigers live in American households than exist in the wild. But the USA is not alone in its love affair with wild animals. As incomes rise elsewhere in the world, so too does the trading and keeping of exotic animals as pets and entertainment, particularly in Europe and Asia.[34]

It's extremely easy to purchase a wild animal. Perusing the Internet will reveal chat rooms, auction sites and dealer sites where one can find a Noah's Ark of animals for sale.[35] Looking for a baby python? They come cheap. At the time of writing, a baby python can be found online for about US$25. Bearded dragons can be purchased for $75. And with a little more ready cash, baby marmosets and capuchin monkeys are available for as little as $350 and tigers for $1000. Swap meets, newspaper ads, fairs and pet stores are other sources.

It's hardly surprising to learn that animals from the wild are inherently unsafe to have around humans. No one would suggest that it is a good idea to unwittingly approach an elephant roaming free in the African savannah. So why do we abandon this common wisdom when it comes to elephants and other wild animals in zoos, circuses and private households? Additionally, although it is difficult to establish a direct causal relationship, the way we treat wild animals may be, at least in part, responsible for the thousands of attacks on humans each year. Most animals captured from the wild to be used in entrainment or in private households are captured as infants. In order to capture wild baby animals, their family members are frequently killed, often in full view of the young.[36] For social creatures such as NHPs and elephants, witnessing the killing of their parents and other family members, followed by separation from their natural environment, can be extremely traumatizing and may cause a lasting impact on an animal's behavior, which can endanger humans. Removed from their natural habitats, a captured animal may be passed from one broker to another several times before being shipped, regionally or internationally, and experiencing extremely cruel transport conditions.

As bad as a captured animal from the wild has it, animals raised in captivity aren't necessarily treated any better, nor are they safer for us. Although NHPs are banned from importation into the USA for the pet trade, they are widely bred there.[37] Charla Nash was mauled by her friend's pet chimpanzee, Travis, on February 16, 2009.[38] He attacked Nash without apparent provocation, ripping off her hands, nose, lips and eyes before fleeing. He was later shot to death. Nash survived, but she is now blind and severely disfigured. Travis had been surrounded by humans his entire life. He was born at the Missouri Primate Foundation, the largest chimpanzee-breeding compound in the USA.[39] The compound, in addition to breeding and selling chimps, rented them out to parties.[40]

No matter the circumstances, by their very nature, non-domesticated animals are dangerously unpredictable. A large part of this unpredictability may be due to how we treat them. Regardless of whether they were taken from their natural habitats or bred in captivity, wild animals kept as pets and in private collections are usually subject to woefully inadequate care, neglect or outright abuse. Most buyers know little about the animals' needs and care requirements, and they are unprepared or unable to provide for them. Birds are most often housed in small cages, depriving them of the very thing that defines them: flight.[41] Caged birds routinely display abnormal behaviors, such as self-mutilation and stereotypies (repetitive movements, such as pacing). NHPs and birds—social animals by nature—may live their entire lives separated from others of their kind and in unnatural, small and bare environments.[42]

Perhaps more disturbing is the ease with which buyers dispose of their exotic pets. One morning in July 2010, a custodian at a Boston high school was cleaning the school lockers.[43] The school was closed for the summer. Imagine the custodian's surprise when a hissing, very irate python fell out of one the lockers onto his feet! A student had abandoned the python in his locker. Several years before the python was found in the locker, a parrot was found abandoned in a nearby automated teller machine. People frequently buy baby snakes, tiger kittens or other animals, only to learn that they quickly grow into extremely large and/or difficult-to-handle adults.[44] Many of those 'cute' baby pythons will quickly grow to 15 feet in length. As the growing animals become unmanageable, they are often kept in chains or small pens and may be beaten into submission. Many are eventually abandoned or left to languish in cages, or die due to physical abuse or inadequate care.[45]

Despite many people's perceptions to the contrary, animals kept for public display or entertainment are often treated no better than those kept in private households. Returning to the two elephants for a moment, it was previously mentioned that Lota's experiences in zoos and circuses are anything but exceptional. Ten years after Tyke's death (she was the elephant who rampaged), John Cuneo Jr., the owner of Hawthorn Corporation, was charged by the USDA with a litany of animal welfare violations, including serious charges, such as mishandling that caused physical harm, discomfort and trauma to the elephants and that created a risk for both these animals and the public.[46] Cuneo admitted guilt, he was fined US$200,000 and 16 elephants were removed from his care. This admittance came years after repeated incidents of elephant rampages, multiple shipments of sick elephants infected with tuberculosis and cruelty to animals.

Animals captured for zoos, circuses and other forms of entertainment are routinely beaten to tame them into submission and to force them to perform.[47] One rather typical case occurred in 1998 when an elephant dealer captured 30 baby elephants in Botswana and shipped them to his warehouse in South Africa.[48] There, he deprived them of food and water and beat them as part of his training process to prepare them for life in captivity. Seven were sold and shipped to zoos in Europe before the dealer was successfully charged with animal cruelty.[49] The so-called 'greatest show on earth', Ringling Brothers and Barnum and Bailey Circus, has been abusing animals for years. Former employees have accused the circus of whipping elephants and beating them with bullhooks, and chaining them for days at a time.[50] USDA inspection reports have repeatedly cited zoos and circuses, including Ringling Brothers, for the deplorable conditions in which they house animals, lack of veterinary care for sick animals, inadequate testing for tuberculosis, lack of vaccinations and physical abuse of animals.[51] To make matters worse, many of the reports cite inadequate safety barriers between animals and the public. All of these factors place the public at risk of not only injury but also infectious diseases. Indeed, some citations were for the actual harms caused to the public. Despite these violations, however, most of these zoos and circuses are still in business.

The typical life of an animal in a traveling zoo or circus involves being kept for months in a small, barren cage, tightly chained for many hours every day, while being trucked from one venue to another.[52] Roadside and non-traveling zoos, particularly the smaller ones with poorly trained staff and inadequate budgets, are notorious for keeping animals in deplorable conditions.[53] Animals are often found lying in their own filth with little food and water. Social animals may be kept in isolation and other animals may be crammed into cages. Even those lucky enough to be in better zoos may eventually end up in the hands of entertainment companies and private collectors if they are deemed as zoo surplus or 'difficult to handle'.

Some, primarily those who profit from using animals as entertainment, contend that such zoos, circuses and other entertainment venues provide education about animals and teach people to appreciate them.[54] Yet there is little evidence to support such claims and, arguably, the opposite is true. Watching an elephant perform handstands or tigers jumping through flaming rings of fire hardly seems to confer respect for animals. Video footage of animals in their natural environments reveals much more about them and helps foster understanding and appreciation of who they truly are.

It is evident that private collectors and entertainment venues are putting the public at peril. One lawsuit (among many) brought against Hawthorn Corp. was on behalf of plaintiffs, many of whom were children, who suffered psychologically as a result of witnessing Tyke's killings.[55] This suit was settled out of court. In 2003 a Bengali tiger named Montecore sank his teeth into Roy Horn's neck (of the famous 'Siegfried and Roy' animal act) and dragged him off stage in front of a horrified audience in Las Vegas. The USDA stated in its final investigation report that the show failed to protect the audience because it had no barrier separating the animals from the crowd.[56] 'The big cats could have easily jumped off the stage and into the audience,' said USDA official Robert M Gibbens who had attended an earlier performance. A chimpanzee named Suzy (the mother of Travis, the chimpanzee who attacked Charla Nash) escaped from the Missouri Primate Foundation compound with two other chimpanzees and was gunned down by a teenager after they turned on him.[57]

Behavioral problems and physical illnesses are common among all wild animals kept as pets or used for entertainment. These issues are primarily a result of their living conditions, malnutrition, abuse and stress.[58] Given these conditions, it's not surprising that animal attacks occur. Even though it is difficult to directly connect our treatment of animals with these attacks, we can establish with little doubt that just keeping wild animals as pets and entertainers is incredibly risky. Primatologist Frans de Waal, after being asked why Travis may have become violent, had this to say:

> A chimp in your home is like a time bomb. [He] may go off for a reason that we may never understand... Usually these animals end up in a cage... even if a chimp were not dangerous, you have to wonder if the chimp is happy in a human household environment.[59]

The bottom line is that no matter how much we think we know a wild animal, and no matter how much we think we are providing an adequate environment, we probably don't and can't.

Of course, even animals domesticated over eons can, and do, attack humans. Often this occurs because of direct provocation or because those animals were specifically trained to be aggressive. But, over thousands of years, we have come to understand the needs of and behavior of domesticated animals far better than those of other animals. In addition they have learned to live with us. This greater understanding on both sides helps to minimize harm. This is not the case with wild animals. With few exceptions, we don't know how to, don't care to or simply can't provide appropriate, rich and safe environments for wild animals.

Admittedly, the total number of human injuries and deaths caused by wild animals kept in captivity is relatively low in comparison with other causes of injuries, such as motor vehicle collisions. However, and more importantly, the animals' poor health that results from our maltreatment leads to another, even more significant public health problem. An even greater danger lurks behind the wildlife trade; one that not just impacts individuals or small groups but threatens the entire globe: infectious diseases.

The rise in infectious diseases

In the past few decades the world has witnessed an unprecedented surge in emerging infectious diseases (EIDs), such as AIDS, SARS, Ebola, 2009 H1N1 (commonly referred to as 2009 swine flu) and H5N1 (or avian influenza). EIDs are defined as the emergence of new or previously unrecognized infectious diseases, the resurgence of a previously known disease in a given place or population, or the emergence of a known disease in a new population.[60] In the past three decades alone we have seen a resurgence of a number of long-known infectious diseases, such as malaria, tuberculosis and cholera, in regions where they were thought to have been successfully eradicated. We have witnessed the emergence of infectious disease agents in novel places, such as West Nile virus for the first time in the USA in 1999. And we have discovered new infectious agents, such as the Nipah virus and the severe acute respiratory syndrome (SARS) coronavirus.[61] Since 1980, more than 35 new infectious diseases have emerged in the human population and 87 human pathogens have been discovered—that's an average rate of about three new pathogens each year.[62]

Why are we seeing such a rise in infectious diseases? Several factors are to blame. First, some of the diseases may not be new at all; they could have circulated among humans for centuries. Yet they are being identified for the first time because of increased surveillance, reporting and modern laboratory diagnostic techniques. While this is certainly true in some cases, Jones et al. found that even after controlling for increased surveillance and reporting, there has still been a significant increase in EIDs in recent times.[63] Their study supports other reports that infectious diseases are indeed on the rise and are becoming an increasing public health threat. Second, the human population is exploding throughout the world—the current population count puts it at about 7 billion and is estimated to increase to almost 10 billion by 2050.[64] As our population grows, available land shrinks and more and more people are forced to live in crowded, urbanized environments, a situation ripe for the easy spread and emergence of infectious agents. Third, HIV/AIDS has enabled a spike in opportunistic infections, which would otherwise occur at very low rates in healthy populations. Fourth, increasing antimicrobial resistance is spurring the development of 'superbugs'. Fifth, humans are traveling around the globe as never before. Our travels significantly increase our chances of catching a disease in one area and unwittingly transporting the infectious agent to another area, where it was never before seen and where little or no immunity exists.

The sixth and seventh factors are climate change and natural habitat loss, which contribute to the rise and spread of several notable infectious diseases. This is particularly evident with vector-borne diseases. These are diseases that are transmitted to humans and other animals by insects and other arthropods, such as mosquitoes, spiders and ticks. A vector's life cycle greatly depends on climatic factors.[65] It turns out that climate change is helping certain vectors to flourish. As a result, some significant infectious diseases are on the rise. Deforestation, land use changes and rapidly changing weather patterns resulting in high rainfall or drought are believed to be causing surges in malaria, dengue fever and Buruli ulcer disease.[66] Cleared land collects rainwater better than rainforests, providing more suitable breeding grounds for malaria-transmitting mosquitoes.[67] Deforestation favors the growth of *Schistosoma*, which is a parasitic worm.[68]

Rising temperatures are at least partially contributing to the rise in outbreaks of mosquito-borne diseases, such as malaria, yellow fever and Saint Louis encephalitis.[69] By shortening the incubation time of these viruses within mosquitoes, accelerating the maturation of mosquito larvae and increasing the feeding frequency of adult mosquitoes, warmer temperatures favor the transmission of the viruses. As will be described later, our encroachment upon and fragmentation of woodland habitats in the northeastern USA is implicated in the rise of Lyme disease.[70] The Nipah virus is a newly discovered pathogen that is causing considerable public health concern because of its ability to infect a broad range of animals and its high lethality among humans.[71] It was first detected in a Malaysian village, where it caused severe encephalitis and high mortality in humans. Habitat loss is believed to have caused a mass exodus of Nipah virus-carrying *Pteropus* 'flying fox' (or fruit bats) as they searched for food.[72] This led the bats to cultivated fruit farms that were planted next to pig farms to allow for the use of the pig manure as crop fertilizers. Unfortunately, the pigs were highly susceptible to the Nipah virus and passed it on to humans.

While these seven factors do indeed contribute to the rise in EIDS, an eighth is rapidly gaining in importance and may be paramount: the global and regional trade in, and production of, animals. As our demand for animals for food, skins, fur and entertainment increases, so does our risk of infectious diseases. Chapter 4 will explore how changes in animal agriculture are contributing to EIDs. The current chapter will focus on the wildlife trade, which is significantly increasing the potential for human contact with existing and, most importantly, novel zoonotic pathogens. These are infectious disease agents that jump from other animals to humans (or vice versa). Most of the known human pathogens are classified as zoonotic.[73] And, of the 175 human pathogens (bacteria, parasites and viruses) that have been classified as emerging or re-emerging, three-fourths come from non-human animals. A 2005 editorial in *The Lancet* proclaimed: 'all new infectious diseases of human beings to emerge in the past 20 years have had an animal source'.[74]

Bushmeat, HIV and Ebola

Arguably, most of the potential pathogens roaming the globe have not yet been encountered by humans. 'For every virus that we know about, there are hundreds that we don't know anything about,' said Dr Dan Bausch, a professor at the Tulane School of Public Health and Tropical Medicine, who studies the Marburg and Ebola viruses and other emerging pathogens in Africa.[75] 'Most of them, we probably don't even know that they're out there.' But as we move deeper and deeper into forests, savannahs and jungles to seize animals for the trade, we risk exposure to exotic insects and animals that may carry novel infectious agents. If the situation is right (or wrong, depending on how we look at it), those pathogens can pass into the human population and spread like wildfire. HIV is a perfect example. Before 1981, scientists never knew such a virus existed. While no one knows the exact sequence of events that led to the first human HIV infection, there is substantial evidence to suggest that it was contact with NHPs through the bushmeat trade that started what is now one of the most significant and devastating pandemics we have ever experienced.[76] More than 65 million people have been infected with HIV and more than 25 million have died.[77]

Worldwide, AIDS is the leading cause of premature death among people aged 15–59 years.

Bushmeat traditionally refers to animal meat derived from the African 'bush' or forests, particularly in Western and Central Africa. Animals typically taken for meat include chimpanzees, gorillas, other NHPs, reptiles, antelopes, rats and bush pigs. However, the bushmeat trade is not restricted to Africa; it is spreading throughout Asia and South America.[78] And, recently, shipments of wild animals for meat have been entering US and European ports.[79] In addition to the bushmeat trade, countless wild animals are shipped regionally and worldwide to supply the growing demand for ornamentation, hunting trophies and traditional medicines. Medicinal products are often derived from the body parts of animals, including endangered species.[80] Examples include bile derived from bears for cardiac illnesses, tiger bones for arthritis and pain, tiger penises for impotency and geckos for diabetes.[81]

Although indigenous groups have been living off wild animals for food, ornamentation and traditional medicines for centuries, several recent changes have led to an unprecedented surge that has become far from sustainable and is dangerous to the health of all.[82] Rapid population growth and increased numbers of wealthy populations in Africa, Asia and elsewhere are creating a strong urban demand for such products.[83] Additionally, logging industries have helped transform bushmeat hunting into a commercial operation and have increased both demand for and access to bushmeat. The logging industries bring roads, trucks, hungry workers, their families and hunters into forested areas that were once inaccessible.[84]

As a result of these combined factors, the trade in wild animal products has become a very profitable enterprise: hunting is cheap and the market price of wild animal products is high.[85] In 2008, the sale of African bushmeat alone was a US $15 billion industry.[86] The economic incentive for bushmeat has shifted hunting from a subsidence activity to large-scale commercial enterprises, often by para-militarized groups, with far greater numbers of animals killed than ever before.[87] Bowen-Jones et al. described the shift aptly: 'money rather than food is now often the prime motivation for hunters'.[88]

The increased hunting of wild animals is creating a ripe opportunity for new pathogens to enter the human population. Virus hunter Nathan Wolfe described how pathogens might jump from monkeys to chimpanzees. In a Ugandan forest, Wolfe and his colleagues witnessed a group of chimpanzees feasting on a freshly killed monkey:

Any disease-causing agent present in that monkey now had the ideal conditions under which to enter a new type of host: the chimps were handling and consuming fresh organs; their hands were covered with blood, saliva and feces, all of which can carry pathogens; blood and other fluids splattered into their eyes and noses. Any sores or cuts on the hunters' bodies could provide a bug with direct entry into the bloodstream.[89]

Replace chimpanzees with human hunters in this scenario and we can see how viruses can easily jump from NHPs and other animals to humans. As hunters and their families butcher and prepare animals for food, many opportunities for pathogen entry exist. All methods of killing animals can expose humans to novel pathogens. Animals captured for the trade, whether for meat, medicines or ornamentation, are routinely poisoned, snared or bludgeoned to death for their bodies or body parts.[90] Each year, thousands of animals are kept in cramped cages for a brief time before being skinned alive or killed by gassing, stomping, electrocution or strangling for their fur or skin.[91] While being killed in these ways, animals frequently soil themselves and release other bodily fluids, which may harbor zoonotic pathogens.

In his article for *Mail Online*, reporter Tom Rawstorne described the killing of one python for his skin in a slaughterhouse in the Indonesian jungle:

The snake is stunned with a blow to the head from the back of a machete and a hose pipe expertly forced between its jaws. Next, the water is turned on and the reptile fills up, swelling like a balloon. It will be left like that for ten minutes or so, a leather cord tied around its neck to prevent the liquid escaping. Then its head is impaled on a meat hook, a couple of quick incisions follow, and the now-loosened skin peeled off with a series of brutal tugs—much like a rubber glove from a hand. From there the skin will be sent to a tannery before being turned into luxury shoes or handbags.[92]

At open markets, where animals are sold live and then butchered on the premises, it is common practice to kill animals by skinning and disemboweling them alive.[93] These methods of slaughter are messy (not to mention horrendously inhumane) and provide many routes by which a pathogen might infect the person doing the killing. Whatever method is employed, every time we handle and slaughter a wild animal there is an opportunity for a new pathogen to enter the human race.

In 1999 a team of investigators reported that they isolated the Simian Immuonodefiency Virus (SIV) in a subspecies of chimpanzees, *Pan troglodytes troglodytes,* in central Africa.[94] Through molecular analysis, they found that the SIVcpz strain closely matched HIV-1, the predominant HIV strain that infects humans. The investigators concluded that SIVcpz is the precursor to HIV and that this subspecies of chimpanzee is the natural reservoir and source of at least three independent virus cross-species transmission occurrences from chimpanzees to humans. We were most likely first infected by exposure to contaminated animal secretions, tissues and blood through hunting, butchering and/or consuming infected chimpanzees.[95]

More recent studies suggest that human exposure to SIV is ongoing and that cross-species transmission occurs more frequently than previously thought. Peeters et al. took blood samples from 573 freshly butchered monkeys sold in bushmeat markets in Cameroon and nearby areas.[96] They also tested blood from 215 wild monkeys kept as pets. They found that almost one in five monkeys sold for bushmeat and more than one in ten sold as pets showed evidence of infection with different SIV strains. They also found that many more species of NHPs than previously thought are infected with SIV and that there are many SIV subtypes. A study of people in 17 villages in Cameroon found that exposure to NHPs is substantial and is not confined to hunters but also occurs among their family members and many others who butcher animal carcasses and prepare the meat.[97]

One of the features of viruses and bacteria that is especially difficult to combat is their ability to adapt to new environments through rapid genetic mutations. The alarming rise in antibiotic-resistant bacteria is a result of bacteria adapting to and surviving antibiotics through genetic mutation. The greater the genetic variability in a species, the greater the chance that some individuals will survive an environmental—or, in this case, pharmacologic—assault. Those strains of bacteria with the right genes to help them survive an antibiotic will pass on their genetic makeup to successive generations, which will then flourish and become the bane of doctors and hospitals everywhere.

When it comes to the ability to survive, of all life forms on earth, viruses are probably about as perfect as it gets. Actually there is a debate as to whether or not viruses should be classified as life forms. Either way, viruses are master replicators. In 24 hours, one virion (a single virus particle) replicates to become ten billion virions—that's a replication rate of almost 116,000 per second! This high replication rate makes viruses especially good at mutation and adaptation. According to the National Institutes of Health medical epidemiologist David Morens,

> when you look at the relationship between bugs and humans, the more important thing to look at is the bug. When an enterovirus like polio goes through the human gastrointestinal tract in three days, its genome mutates about two percent. That level of mutation—two percent of the genome—has taken the human species eight million years to accomplish. So who's going to adapt to whom?[98]

Unfortunately for us, three-fourths of all new human pathogens that have emerged since 1980 are viruses.[99] This forecasts a very troubling future. As geneticist and Nobel prize winner in medicine Joshua Lederberg says, 'the single biggest threat to man's continued dominance on the planet is a virus'.[100]

Human T-lymphotrophic viruses (HTLV) cause several types of adult leukemia and neurological illnesses in humans. Like, HIV, HTLV is a retrovirus. These are transmitted through blood and body secretions. Retroviruses insert themselves into the hosts' (in this case, humans') DNA through a process that allows them to frequently mutate and to continuously evolve and adapt to new environments. Their high mutation rate and their ability to hide inside the host's own cells make them extremely difficult to eliminate, as is evident with our experience with HIV.

HTLV originated from the simian T-lymphotrophic viruses.[101] A recent investigation of 11 rural villages in Cameroon by Nathan Wolfe and his team found that of 200 people interviewed, almost 40 percent reported exposure to NHP blood and secretions, mainly through hunting and butchering.[102] Blood samples from the villagers revealed widespread infection with HTLV, as well as two previously unknown retroviruses. In another study, Wolfe and his team found evidence of Simian Foamy Virus (SFV) infection in individuals in Cameroon.[103] This is another retrovirus of NHP origin and is widespread among African NHPs. SFV has also been found, not infrequently, in blood samples from laboratory and zoo workers exposed to NHPs.[104]

Because of our close genetic relationship with other primates, we are often especially vulnerable to contracting the pathogens they carry. Cameroon, and indeed all of Central Africa, is home to some of the highest densities of NHPs and is considered a hot spot for a host of new zoonotic diseases.[105] Already we have witnessed, from Central Africa alone, the emergence of some notable zoonotic pathogens, including Marburg, Ebola, monkeypox, HIV, HTLV and SFV. A novel poxvirus, a member of the same viral family as smallpox, has recently been discovered in red colubus monkeys in western Uganda.[106] Although how some of these new viruses, such as SFV and the new poxvirus, affect humans has yet to be determined, others have proved to be highly lethal in humans.

Emerging infections are coming from non-primate animals, too. The Ebola virus causes hemorrhagic fever and is one of the most lethal pathogens affecting humans. Depending on the viral strain, the death rate among those who contract Ebola varies from 50 to almost 90 percent.[107] Although human outbreaks of Ebola have been traced back to contact with chimpanzees and gorillas, the primary reservoir of the Ebola virus is suspected to be the fruit bat.[108] Other animals found to be infected with Ebola include forest antelopes, rodents and shrews in Central Africa.[109]

Taken together, these studies suggest that a considerable proportion of NHPs and other animals in Central Africa are infected with a wide range of viruses (including many that have yet to be detected) and that transmission of these viruses into the human population is significant and actively ongoing.[110] To make matters worse, HIV/AIDS is highly prevalent in Central Africa. The pandemic has left a large immunocompromised human population that is extremely susceptible to new infections.[111] A recent study found that among 191 HIV-infected people in Cameroon, 80 percent butchered wild animals, 84 percent consumed NHPs and more than 8 percent kept NHPs as pets.[112] Humans and all other animals are like nightclubs for viruses. In us, different viruses can exchange greetings, mingle and swap genes to create new viral strains. This hazard is elevated in those with pre-existing infectious diseases. In HIV-infected individuals, newly introduced viruses from other animals and circulating HIV can potentially recombine, creating new zoonoses that may be even deadlier.[113]

Other pathogens

The recent emergence of zoonotic infections is not confined to Central Africa. The Nipah virus from Malaysia is one example. It shares similarities with another paramoxyvirus: the Hendra virus. Hendra was identified in 1994 in Australia when two of three infected people died after contact with horses suffering from a severe respiratory disease.[114] It is believed that the horses were infected by fruit bats. Of interest, however, is that there was no evidence of Hendra virus infection in a study of wildlife rehabilitators who frequently handled these bats.[115] This suggests that some intermediate host may have been required for the amplification in and/or adaptation of the virus to the bats.[116] Nipah and Hendra viruses may not be new. Phylogenetic studies suggest that they have been around for a long time but only caught our attention after ecological changes led to human contact with infected animals.[117]

SARS, which caused a near-pandemic and resulted in more than 8000 cases and 774 deaths, likely had its start in the bushmeat trade in the Guangdong Province in China and nearby regions.[118] Although thousands of civets—small arboreal mammals exploited for their musk-producing glands—were slaughtered en masse because they were suspected to be the source of the infection, SARS is now believed to have emerged from infected bats.[119] The bats were captured from the wild and traded for the live markets of China.

In addition to occurring throughout Southeast Asia, live markets are increasing in New York and California due to the growing Asian immigrant populations in both states.[120] Regardless of their location, these markets are miserable places for animals and provide an opportunity for the spread of diseases. Here, live animals of a wide variety of species sold for food are crammed into cages where they are unable to move, often causing those at the bottom to be crushed to death.[121] The animals are often deprived of food, water and shelter and are exposed to extreme heat or cold. The slaughtering methods used are also often extremely inhumane.[122] Turtles have their intestines removed while they are still alive and live birds are placed in plastic bags until they suffocate.[123] In Chinese markets, cats and dogs sold for food may be slowly bled to death or bludgeoned.[124] Hygiene in the markets is extremely poor, with the animals shedding copious amounts of feces, urine and other excretions.[125] These secretions may contain large numbers of pathogens that are potentially hazardous to humans. Because of the openness of these markets, newly introduced animals may come into direct contact with sales clerks and customers, in addition to the animal handlers and butchers.

After many studies examined how SARS appeared and spread within these markets, researchers now suspect that at some point in the wildlife supply chain, infected bats were brought into contact with susceptible hosts, such as civets, in whom the virus amplified.[126] The intermingling of species established a cycle in which susceptible animals and humans could become infected. We may learn next that bats were not the primary source of SARS after all and that a yet-unidentified creature was involved—one that may have infected both the bats and the civets.[127]

Virologist Ron Fouchier of Erasmus MC University in the Netherlands says that regardless of whether or not bats were the primary source, it would be a mistake to wipe them out. The problem is not the bats, he says, but rather what humans do with them. People eat bat meat and use their feces in medicine, he says. 'Rather than blaming animals and killing them, we should change our behavior.'[128] After the SARS epidemics subsided, an editorial in the *American Journal of Public Health* made this observation: 'The concentration of animals, their overlapping sojourns in the markets (allowing disease to spread through vast numbers of animals), and their interactions with humans (facilitating human infection) make these markets ripe for zoonoses. Once an epidemic starts among animals, it can spread to animals reared in less cruel conditions.'[129]

Some of the pathogens that have emerged in the past few decades, such as Ebola, have remained largely confined to localized populations. However, as we ship animals regionally and across the globe, we risk dispersing zoonoses worldwide. The only time Ebola was known to have ever entered the USA was when infected NHPs were imported for biomedical research.[130] Populations participating in the bushmeat trade are increasingly connected with urban communities, which facilitate long-distance transport of bushmeat.[131] SARS became a near-worldwide pandemic, in part, because infected animals were probably shipped throughout Southeast Asia.[132] The Nipah virus could have remained confined to the Sungai Nipah New Village where it began. Instead, it spread throughout Malaysia and Singapore through the trucking of infected pigs.[133] The Marburg virus, a cousin of Ebola, is originally from Africa. Yet it was first detected in 1967 in the German town of Marburg after laboratory workers caught it from infected African green monkeys imported from Uganda.[134] In the USA, rabies was introduced to the mid-Atlantic states in the 1970s when raccoons captured from rabies-endemic areas were used to repopulate hunting pens.[135] As James Hughes, longtime director of the National Center for Infectious Diseases at the Centers for Disease Control and Prevention (CDC), quipped in 2003, 'People have looked very hard for the source in nature of Ebola virus, and they haven't found it...I certainly don't want to find it as the result of the importation of an infected animal.'[136]

A pathogen's ideal environment

If viruses, bacteria and parasites could tell us about their ideal environments, we would hear them describing the animal trade among their top choices. We have seen a glimpse of what happens to animals once they reach their intended destination. But what occurs before and during transport is no less harmful and, as a result, these animals are highly prone to catching and transmitting infectious agents. The unfortunate reality is that few laws exist to protect animals from harm during any phase of the trade.[137] When regulations exist, they are rarely enforced or the penalties are so minor that they provide almost no deterrent.[138]

In fact, the paucity of regulations has spawned a massive underground, illegal wildlife trade in addition to the legal trade. The illegal trade is highly lucrative. In comparison with other illegally traded items, such as guns and drugs, animals are quite cheap to come by, the risks of penalties are drastically lower and the payoffs can be much greater.[139] For example, ground rhinoceros horn can earn higher profits than the equivalent in gold or cocaine.[140] After the smuggling of drugs, the illegal wildlife trade is the most valuable illegal commerce in the world—even more profitable than the smuggling of weapons or humans.[141] Like other illegal trades, the illegal wildlife trade has become an increasingly well-organized endeavor with worldwide criminal syndicates creating black markets and smuggling routes.[142] According to the US State Department, 2–5 million birds, including parrots, eagles and hummingbirds, as well as millions of reptiles and mammals, are smuggled worldwide annually.[143] However, accurate estimates are hard to come by because of the illicit nature of the trade. As is the case with the legal trade, the USA, Europe, China and other Asian countries are the greatest consumers of illegal wildlife.[144]

Regardless of whether the trade is legal or not, harsh capture techniques kill many animals before they are ever shipped anywhere.[145] For example, one-third of all captured birds from Tanzania and up to half of finches and waxbills captured in Senegal die before export.[146] During transport, animals are subjected to extreme stress. Virtually all international commercial trade of wildlife occurs by air. The International Air Transport Association has established only voluntary guidelines for shipping animals. Additionally, these are only enforceable for species listed under CITES. Even when CITES-listed species are involved, violators of the guidelines are usually merely given warnings or fined minimal penalties. Thus there is little incentive for wildlife traders to follow the guidelines. To cut shipping costs, traders pack as many animals as possible into flimsy crates or cardboard boxes, sometimes taping the animals together or binding them to restrict their movements.[147] Animals at the bottom of the crates are often smothered and crushed to death.[148] Overcrowding, exposure to extreme temperatures, filthy conditions, poor diets, dehydration and disease are the norm.[149] As a result, a significant number of the animals (60–70 percent of reptiles and birds) die from the transport conditions alone.[150] The animals who do survive the shipping process are in such poor health that many of them die shortly after their arrival at their intended destinations. It is estimated that nine out of ten reptiles who survive shipment into the USA die within their first year of captivity and one in ten birds die within 30 days.[151]

Illegal traffickers are increasingly devising more 'resourceful' and harmful ways to smuggle animals across borders. To restrict their movement or keep them from crying out, animals are bound, gagged and even drugged into unconsciousness and then stuffed into all manners of items.[152] One investigator described how at a market in Ecuador he was offered a parakeet: 'I asked the seller how I would get it on an airplane. "Give it vodka and put it in your pocket," he said. "It will be quiet." '[153] Hummingbirds have been found bound and stuffed into empty packs of cigarettes.[154] A US agent on the US–Mexico border found baby monkeys crammed into a car's air conditioning ducts. Most of them died from suffocation. Animals have been smuggled stuffed into thermoses, stockings, toilet paper tubes and hair curlers.[155] As one Mexican Government wildlife expert reports, 'For every 10 animals trafficked, only one survives.'[156]

Almost all of the same factors that cause animals in the trade distress and suffering also cause immunosuppression, leaving them extremely vulnerable to new infections. As a result, the trade creates very sick animals and conditions ideal for pathogens to multiply. Additionally, of all the determinants contributing to the emergence of zoonotic pathogens (such as ecological factors, natural selection and personal behavior), 'species-jumping' events that expand the range of viable hosts may be among the most important.[157] Holding different populations of animals, particularly sick animals, together during shipment or while housed at pet stores, zoos, circuses, laboratories, markets and hunting pens may result in new strains of pathogens that might not have occurred otherwise.[158] Karesh and Cook aptly summarized the zoonotic risk: 'Daily, wild mammals, birds and reptiles flow through trading centers where they are in contact with humans and dozens of other species before being shipped to other markets, sold locally, and even freed back into the wild with new potential pathogens.'[159]

One global infectious disease world

In 1980 the World Health Organization (WHO) announced the global eradication of smallpox.[160] This is one of the most devastating human infectious diseases encountered and its eradication was a tremendous public health victory. We were able to eradicate smallpox in large part because the virus infects only humans and no other species serves as a reservoir. But, with the WHO announcement, we may be singing our victory song too soon. In 2003 an outbreak of monkeypox made headline news as it spread across half a dozen states in the Midwestern USA.[161] Epidemiologic investigations confirmed that the disease was introduced into the country when a shipment of infected African Gambian rats were sold to pet dealers, one of whom housed the rats with prairie dogs.[162] The prairie dogs subsequently contracted monkeypox, were then sold as pets and transmitted the infection to 71 people. 'Basically you factored out an ocean and half a continent by moving these animals around and ultimately juxtaposing them in a warehouse or a garage somewhere,' said Jeffrey Davis, chief medical officer and state epidemiologist for infectious diseases at the Wisconsin Division of Public Health.[163]

Monkeypox actually entered the USA for the first time in the 1950s, when several outbreaks occurred in NHPs shipped to laboratories.[164] Interestingly, it was not found in the source NHPs free-living in India and Southeast Asia, suggesting that the NHPs shipped to the USA became infected at some point during their transportation. Monkeypox virus is closely related to the smallpox virus but, thus far, it is not highly lethal in humans. However, with opportunities to jump between species and grab more genetic material, it could evolve into a new pathogen to be reckoned with, similar to smallpox. A far deadlier monkeypox strain than that shipped into the USA through Gambian rats causes a disease that is 'virtually indistinguishable' from typical smallpox, according to virologist Mark Buller of St Louis University.[165] About 10 percent of those affected by this Congo Basin strain die—a rate approximating the African death rate from smallpox. Worse yet, evidence suggests that monkeypox in the Congo Basin is evolving so that it could become easily transmissible from person to person.

Not all zoonotic infections are as media-grabbing as monkeypox. Nevertheless, they can pose significant health burdens and can put the most vulnerable humans at risk from serious illness. In the USA, contact with pet reptiles causes frequent outbreaks of *Salmonella* infection.[166] Approximately 7 percent of all *Salmonella* infections and 11 percent of those among people younger than 21 years are caused by direct or indirect contact with reptiles—about 74,000 each year.[167] Although *Salmonella* usually causes a mild, self-limiting gastroenteritis, young children, the elderly and immunocompromised individuals are at risk of more severe diseases, including meningitis and sepsis, and death.[168] Young children are especially prone to infection from reptiles because of their frequent contact with them at petting zoos, fairs, flea markets and in homes. In 1975 the US Food and Drug Administration banned the import and sale of turtles of less than four inches in carapace size as pets (incidentally, there was no ban on exports).[169] This four-inch rule was intended as a guideline above which it would be difficult for children to put turtles in their mouths like toys. Despite this ban, however, annual outbreaks in the USA continue, in part due to limited enforcement of the ban, but also because a host of other reptiles and amphibians, whose sales are not restricted, are carriers of *Salmonella*.[170]

Salmonella infection in reptiles and amphibians tends to be asymptomatic (i.e. the animals don't show any symptoms) and quite common. Thus reptiles can shed *Salmonella* in their feces over prolonged periods of time with nearby humans being none the wiser. An analysis conducted in Germany and Austria of 48 reptile species found *Salmonella* in 54 percent of the sample.[171] Of these positive individuals, most came from pet stores. A study in Japan found that 74 percent of reptiles from pet stores carried *Salmonella*.[172] Although all reptiles may be carriers, several studies, as will be described later, suggest that *Salmonella* prevalence may be higher in captive reptiles than in free-living animals. Given that captivity can be stressful and that pet stores are notorious for inhumane and unhygienic conditions, it is reasonable to postulate that reptiles in such stores are more likely to carry and/or shed pathogens such as *Salmonella* than free-living animals.

Humans become infected not only by direct contact with reptiles and amphibians but also by contact with their environments, such as aquarium water.[173] In 1996 a *Salmonella* outbreak occurred among 65 children after they attended a Komodo dragon exhibit at a metropolitan zoo.[174] None of the infected children had touched the dragon but almost 83 percent had touched the wooden pen in which the animal was housed. Other *Salmonella* infections and outbreaks in the USA have occurred after indirect or direct contact with African dwarf frogs, boas, iguanas and bearded dragons.[175] In Europe and elsewhere, reports of human *Salmonella* infections are paralleling the rise in importation of reptiles and amphibians.[176]

Salmonella is, of course, just one of many infections that have passed from wild animals to humans. NHPs carry herpes B virus, mongooses carry cowpox, parakeets carry *Chlamydia*, hedgehogs carry *Yersinia* and hamsters carry tularemia.[177] In one study, 17 out of 28 different species of squirrels, gerbils, mice and chipmunks purchased from trading companies in eight countries were found to be infected with *Bartonella* bacteria, including six novel bacteria species.[178] *Bartonella* can cause inflammation of the heart and the central nervous system and trench fever in humans. An outbreak of psittacosis, a parasitic infection that causes respiratory illness, occurred among people who purchased birds from nine US pet stores in 1995.[179]

Exposure to animals at circuses and zoos can also result in human infection. At an exotic animal farm in Illinois, 12 circus elephant handlers showed evidence of infection after 3 elephants died of tuberculosis.[180] A recent study found that *Blastocystis*, a parasite that causes gastrointestinal illness, is spreading among animals in zoos and transmission is occurring between animals and zookeepers.[181] Seven animal handlers tested positive for tuberculosis after an outbreak occurred among monkeys and rhinoceroses at a zoo in Louisiana.[182] As one study author attested, 'Zoos are indeed a hot spot for interspecies spread of infectious diseases.'[183] The same can probably be said for circuses and animal amusement parks. Between 1990 and 2000 in the USA, more than 25 outbreaks of human infectious diseases occurred due to animal exhibitions alone.[184]

Pavlin et al. looked at the types of mammals imported into the USA between 2000 and 2005 and assessed their potential to transmit 27 different zoonotic diseases.[185] The investigators found that the imported animals were capable of carrying a myriad of significant infectious agents and diseases, including Marburg virus, Ebola virus, herpes B virus, rabies, tuberculosis, avian influenza (H5N1), yellow fever, tularemia and anthrax. Thanks to the wildlife trade, we are unwittingly shipping many of these pathogens and diseases throughout the world. Michael Osterholm, director of the Center for Infectious Disease Research and Policy at the University of Minnesota, fittingly described the public health risks as a result of the trade when he stated, 'We now have this potential to make it literally one global infectious disease world.'[186]

Suburban monkeys and the loss of biodiversity

We are creating a global infectious disease world in more ways than one. In addition to directly increasing our risk of epidemics, the wildlife trade is causing devastating destruction to our ecosystems and loss of biodiversity. As will be explored later, loss of biodiversity has already caused a rise in some very notable infectious diseases. Worldwide, approximately 1.8 million species of animals, plants, insects and other life forms have been identified.[187] But that is a very small number compared with how many we are not aware of. Estimates of the true number of species on earth range anywhere from 2 to 100 million (the majority being microbes). That's a wide range, but it shows us just how little we know about the spectrum of life on our planet. Sadly, if trends don't reverse, we will probably never know just how rich our earth's biodiversity is... or once was.

Almost universally, the fate of wildlife populations is grim. A massive extinction of animals is taking place.[188] At a recent UN conference in Nagoya, Japan, scientists pointed out that the earth is losing species at 100–1000 times the historical average.[189] According to the International Union for the Conservation of Nature Red List of threatened species, more than one-fifth of all the currently known vertebrate species are threatened today.[190] Over 100 species of amphibians are estimated to have become extinct since 1980 and of the remainder, one-third are under threat.[191] Almost 100,000 tigers existed worldwide about a century ago but today fewer than 3500–5000 may remain in the wild.[192] Of the 145 species of parrots in the Americas, almost one-third face extinction.[193] Our closest living relatives, the great apes, are on the brink of extinction. Almost half of the 634 species of NHPs may soon vanish.[194] More than 70 percent of Asian NHPs and more than 90 percent of NHPs in Vietnam and Cambodia are threatened.[195] Even previously threatened species, such as elephants, which due to conservation efforts were experiencing recovery, may now be at serious risk again.[196]

Eminent paleontologist Richard Leakey refers to the current biodiversity crisis as the sixth great extinction.[197] The last occurred 65 million years ago at the end of the Cretaceous period. It led to the fall of the dinosaurs and the ascendancy of humans and other mammals. Leaky estimates that yearly between 17,000 and 100,000 species vanish completely. 'For the sake of argument,' he says, 'let's assume the number is 50,000 a year. Whatever way you look at it, we're destroying the Earth at a rate comparable with the impact of a giant asteroid slamming into the planet, or even a shower of vast heavenly bodies.'[198] By Leakey's estimates, half of the world's species will become extinct within the next century and most biologists polled in the USA are convinced that a mass extinction of plants and animals is underway.[199] What makes the sixth extinction so unlike the five prior ones is that the cause is almost entirely human. And, unlike the last great extinction, this one is not likely to benefit us.

The human-derived causes of the loss of biodiversity are multifactorial. As the human population grows, so does the demand for land and other resources.[200] Perhaps the greatest threat to species survival is habitat loss.[201] Deforestation and other habitat loss is increasingly occurring due to the conversion of land for intensive livestock and agriculture, logging and to make room for the ever-increasing human population.[202] While habitat loss may be the main cause of extinction overall, the wildlife trade is also playing a major role, and for many species and in many regions of the world the wildlife trade is the most immediate threat to species' survival.[203] For example, in the Congo Basin, commercial hunting of wild animals for meat has already caused numerous local extinctions throughout the region.[204] According to Dr John Behler of the Wildlife Conservation Society, the trade for food and traditional medicine is causing the demise of turtle populations.[205] Some 50 parrot species are in jeopardy due to the exotic pet trade. Prior to the CITES ban on trade in ivory, the number of African elephants fell by half in ten years, and they were at risk again when both Tanzania and Zambia proposed re-opening the ivory trade.[206] Fortunately these bids, backed by China and Japan, were rejected by CITES in March 2010. The wildlife trade and habitat destruction are the biggest threats to NHPs.[207] In all, the wildlife trade threatens about one-third of all mammals and birds.[208]

The trade in wildlife continues even as species diminish in number. In fact, the more rare the species, the greater the public demand and economic incentive. This creates a positive feedback loop that leads to even greater exploitation.[209] Additionally, as a species in one area is exploited to extinction or near-extinction, traders just either move to other regions or broaden their repertoire to include other species in a never-ending cycle. Vincent Nijman of Oxford Brookes University describes this cycle: 'we see species that are in fashion traded in great numbers until they are wiped out and people can't get them anymore. So another comes in, and then that is wiped out, and then another comes in.'[210]

By removing animals, we risk serious repercussions for the entire local environment.[211] The collecting of animals itself often destroys habitat. For example, trees are commonly felled to capture wild birds, diminishing nesting sites for future generations.[212] Burrows are destroyed to capture snakes and tortoises, again destroying habitats for future generations.[213] Forests and other habitats are burned down to 'out' the targeted animals.[214] Toxins and chemicals, such as gasoline, are used to drive reptiles from their homes. In addition to harming the species captured, the wildlife trade causes a cascade of events that disrupts ecosystems and threatens the survival of other species that are not even part of the trade.

Each species plays an important ecological role in its natural environment.[215] Many of those threatened serve as vital seed-dispersal agents and their removal or diminishment threatens the very survival and diversity of our most ecologically important forests and other ecosystems.[216] The trade removes animals serving as important food sources for other animals.[217] Alternatively, removal of many of the large predators impedes the keeping of other populations in balance.[218] In essence, the loss of a single species can have far-reaching effects and can disrupt the ecological balance of an entire forest.[219]

As if removing ecologically vital species is not enough, the trade is causing another serious worldwide problem that further exacerbates the biodiversity crisis: the introduction of non-native animals that endangers native species.[220] Non-native species are released accidentally, escape or are released intentionally by people who are no longer able, or want, to care for them.[221] Intentional release frequently occurs when people purchase young animals only to find that they grow into an unmanageable size. The introduction of non-native animals threatens native species by competing for resources and habitats, by preying on native species for food and by altering native ecosystems.[222] In Florida, a major importing site in the USA, non-native squirrel monkeys, macaque monkeys, Burmese pythons, South American parrots, African Nile lizards and other exotic animals have established themselves in the Everglades and are now commonly seen roaming neighborhoods.[223] Florida residents are now looking out of the doors of their nice suburban homes and seeing African monkeys for the first time swinging through the trees and rummaging through their trash cans!

In addition to introducing non-native species, the trade in wildlife introduces infectious diseases to new populations.[224] One of the biggest threats to a large number of amphibians is the disease caused by a chytrid fungus, which is believed to have originated in South America and is wiping out whole populations across the globe. It was facilitated by the wildlife trade and its consequent release of non-native species carrying the fungus.[225] Ranaviral disease of amphibians is also believed to be globally spread via the wildlife trade.[226] As we help spread diseases to other species, we also risk a spill-back effect, in which the zoonotic pathogens come back to infect us.[227]

Independently, the wildlife trade is creating what conservationists have dubbed the 'empty forest syndrome'.[228] Progressively, forests and other natural habitats are being emptied of wildlife. Confronted by the combination of habitat loss and fragmentation, climate change, pollution and the wildlife trade, animals are experiencing an assault of alarming magnitude. Magnifying this attack, a significant number of animals involved in the wildlife trade are taken from biodiversity 'hot spots', the ecologically richest and most species-diverse places on earth.[229] Many of these hot spots are crucial carbon sinks and their destruction exacerbates global warming, further perpetuating the cycle of biodiversity loss.[230] As an agent of the US Fish and Wildlife Service stated, 'people don't realize when they buy an exotic pet they are taking the rain forest and putting it in a coffin'.[231] Regardless of whether we are taking wild animals for pets, food or other purposes, we are devastating our ecosystems.

The human impact of biodiversity loss

Not only is this biodiversity loss bad for ecosystems, it's also bad for us. Evidence suggests that the greater the diversity of species, the less the chance that humans will contract zoonotic pathogens. Although it might be intuitive to think that greater diversity of species also means greater opportunities for infectious agents to enter the human population, evidence is suggesting that the opposite may be true. Ostfeld describes several mechanisms by which high biodiversity can buffer against the transmission of pathogens, including the following:

1. Greater species diversity reduces the population of an important natural reservoir (such as an animal species) for pathogens.
2. Greater diversity reduces the population density of pathogen-carrying vectors (e.g. insects).
3. Greater diversity reduces the encounter rates between vectors and reservoirs.[232]

In summary, the greater the diversity in an ecosystem, the less the chance of one reservoir species becoming dominant and, in many cases, the less the chance that humans will encounter a pathogen-carrying vector.

An understanding of the transmission of the pathogen that causes Lyme disease illustrates how species diversity affects a natural reservoir or vector of an infectious agent. A reservoir serves as a host on which an infectious agent depends to survive or multiply, but usually does not actually get sick from the infectious agent or can carry the infectious agent for a long time before getting sick. A vector transmits the infectious agent from the host to another animal, who does get sick. In the case of Lyme disease, an illness caused by the bacterium *Borrelia burgdorferi*, white-footed mice, short-tailed shrews and eastern chipmunks serve as important reservoirs. In the mouse, for example, the bacteria can find a hospitable environment in which to grow and multiply. The vector here is a tick of the genus *Ixodes*, one of which is commonly referred to as the 'deer tick'. A tick contracts *Borrelia* after it feeds off a white-footed mouse carrying the bacteria; the tick then transmits the bacteria to humans (and other animals) when it feeds off them.

The greater the biodiversity in a North American forest, the more competition white-footed mice have for survival and the less likely they are to dominate the forest. Forests with high diversity will include the white footed-mouse, but also a large number of other animals in whom the bacteria don't live and multiply so readily, but who are equally good sources of food for ticks. Thus, the greater the number of uninfected animals, the fewer encounters ticks will have with infected (reservoir) animals and the less likely ticks will carry the bacteria. Therefore, this reduces our risk of encountering a tick carrying the *Borrelia* bacterium. Ostfeld refers to the phenomenon by which high biodiversity reduces infection risk as the 'dilution effect'.[233]

In the USA and elsewhere, Lyme disease is on the rise. The CDC estimates that with approximately 20,000 new cases reported each year, Lyme disease is the most common vector-borne disease in the USA, and the annual rate of reports has more than doubled since 1991.[234] If not caught and treated in time with appropriate antibiotics, Lyme disease can result in serious cardiac and neurological repercussions, including chronic pain and numbness, paralysis and visual problems.

In the northeastern USA, we have extensively fragmented our forests (a euphemism for 'suburbanized'), resulting in biodiversity loss and unfettered population growth of small animal reservoirs of Lyme disease—those animals who are better able to adapt to the sparse forest patches than other mammals.[235] Based on the dilution effect theory, Ostfeld and Keesing hypothesized that human Lyme disease incidence rates would be lower near habitats containing greater diversity.[236] After analyzing state and multistate regions in the USA, they found a significant negative correlation between the species richness of small land mammals and reports of Lyme disease. Hence, the greater the species diversity, the fewer the number of cases.

The dilution effect has been supported by studies of other formidable infectious diseases, including West Nile virus illness, hemorrhagic fevers, leishmaniasis, African trypanosomiasis, Chagas disease and Rocky Mountain spotted fever.[237] Of course, there may be cases when biodiversity loss in fact causes the very reservoirs or vectors of certain pathogens to be reduced, thus decreasing our risk of those infections.[238] This will require further investigation. Regardless, there is ample evidence to suggest that some very serious infectious diseases today are becoming greater threats, in part due to reductions in species diversity.

Are we just crying wolf?

Alarmists can be rather irritating. Either they cry wolf when there is none, or they are right—and thus are even more annoying! No one likes to hear the bad news, but we are truly endangering ourselves if we ignore the disturbing realities of the wildlife trade and don't take action against it. In order to minimize the threat of this trade, we have to ensure that we choose actions that will be effective. It might seem that all that's really needed to prevent the animal trade from causing infectious disease havoc is simply to monitor the shipment of animals. However, given the immense numbers imported annually to each nation, it is almost impossible for border officers to reliably track them. As Gerson et al. stated in their review of inspections in Canada, 'more than 12 million commercial shipments are imported...annually, and only about 2% of these are physically inspected'.[239] In the USA, with fewer than 100 inspectors monitoring nationwide imports, they would have approximately three seconds to inspect each animal if that's all they did every minute of every work day.[240] Even if we were to drastically increase shipment inspections, simple inspections would still not detect pathogens if, as is commonly the case with *Salmonella* and other pathogens, the animals carrying them showed no overt signs of illness.

Additionally, much of the wildlife trade occurs domestically, from region to region within a nation, leaving little to no paper trail and minimal opportunity for inspection.[241] To supply the US pet trade, an estimated 3–5 million reptiles and amphibians are captured from the wild in the state of Louisiana alone.[242] So monitoring of shipments is not in itself a viable solution. Other options offered to prevent the trade from spreading exotic infectious diseases include screening of animals with laboratory tests, pre-emptive treatment for known diseases, and quarantine of animals.[243] As a report by the CDC stated,

> Many of these solutions are not feasible or practical to use on the large volume of animals that are being imported and cannot be employed to prevent new or emerging pathogens or infections. Ultimately, import restrictions may be the only means of preventing introduction of exotic infections.[244]

Breeding wild animals for the commercial trade is now a common practice, partly because of an implicit assumption that it will ease the risk of infectious diseases, prevent the destruction of ecosystems and preserve species. But does breeding do all of this? Many reptiles supplied for the trade in the USA are bred, but they still cause *Salmonella* outbreaks year after year.[245] In the USA, the Wild Bird Conservation Act of 1992 outlawed the import of most wild-caught birds.[246] As a result, most birds purchased at pet stores in the USA are bred in captivity, but they still transmit psittacosis, which can cause a serious pneumonia, to humans.[247] In a study of birds sold at nine different pet stores in Atlanta, more than one in ten people who bought them contracted chlamydiosis and suffered acute respiratory illness.[248] A breeder in Oklahoma supplied the pet stores. Captive-bred NHPs naturally carry herpes B virus, which can cause a deadly inflammation of the spinal cord and brain (encephalomyelitis) in humans.[249]

Breeding farms may actually increase levels of certain infectious diseases. According to the CDC, small turtles sold as pets in the USA frequently come from breeding farms, where the animals are housed in crowded ponds and nesting areas in a way that promotes *Salmonella* transmission.[250] Even though attempts are made to treat turtles, turtle eggs and turtle breeding ponds with antibiotics and other methods, the continual shedding of *Salmonella* by many turtles may be, as the CDC declares, 'stress related'. Several studies of turtle farms report a high prevalence of *Salmonella*.[251] Even more troubling is that the use of antibiotics to prevent *Salmonella* infections in animals is leading to antibiotic-resistant strains. One study of breeding farms in Louisiana, which routinely employ antibiotics, reported the presence of gentamicin-resistant *Salmonella* strains.[252] Other studies have found multiple pathogens resistant to antibiotics on wildlife breeding farms.[253] In contrast, many studies have found that *Salmonella* prevalence is much lower in free-living reptiles in comparison with captive-bred and pet reptiles.[254] The investigators conducting these studies suggest that the stress of captivity renders animals more vulnerable to infection.

Investigators from veterinary schools in Pennsylvania and North Carolina did not detect *Salmonella* in any of the free-living turtles they sampled.[255] Like the CDC, these investigators postulated that 'captive reptiles may be crowded or subjected to poor hygienic protocol', thus increasing their risk of carrying and shedding *Salmonella*. Most recently, a California breeder of African dwarf frogs has been identified as the source of a *Salmonella* outbreak that caused sickness in more than 200 people in the USA.[256] In Chapter 4, we will explore how zoonotic pathogens flourish in intensive farms breeding and housing animals for food. Besides not necessarily being a more humane solution for individual animals, breeding and keeping animals in captivity is not only no guarantee against infections but may actually increase our risk.

Rather than taking the pressure off species, breeding animals appears to stimulate the trade in endangered species. First, many animals don't breed well in captivity or their breeding is cost-prohibitive.[257] Raising a farmed tiger to maturity, for example, is 250 times as expensive as poaching a wild tiger in India.[258] Second, breeding farms are frequently stocked with wild animals.[259] For example, turtle farms in the USA are stocked by capturing adults and eggs from the wild.[260] Indeed, several investigators found that the primary purchasers of wild-caught turtles are turtle farms, suggesting that they are a major threat to wild turtle populations.[261] Third, among those who purchase animals for traditional medicines, ornamentation or food, the majority prefer animals from the wild over farmed animals as the former are considered 'purer'.[262] Thus the trade of captive-bred animals would likely serve as a cover for the trade of wild-caught animals since it is almost impossible for trade agents to distinguish between the two.[263] Fourth, increased farming of exotic animals could also increase the demand for endangered species, spurring the illegal trade.[264] The legal trade, whether of farmed or wild-caught animals, appears to act as a stimulant to the underground trade.

As long as the trade continues—regardless of whether it is legal or illegal, or involves captive-bred or wild-caught animals—it will be problematic. It will threaten ecosystems and species survivability. It will cause immense suffering in animals, and risk human injuries and the worldwide spread of known and novel infectious diseases.

Additionally, some of our most vital medicines, including chemotherapeutic agents, come from the forests. With the destruction of ecosystems, we may lose some of our greatest medical treatments before they are even discovered. The wildlife trade may also play a role in crimes against humanity. The Janjaweed, the militia that has carried out genocidal attacks in Darfur, for example, is slaughtering elephants by the hundreds to earn easy money from the illegal ivory trade.[265] Exotic animals and their body parts have become the new blood diamonds. The illegal animal trade goes hand in hand with transnational crime, including trade in ammunition and narcotics, and the trafficking of people.[266] Terrorist groups may also be engaged in wildlife smuggling to buy ammunition and provide financial support for their activities.[267]

The animal trade is now a major chink in the public health armor. Through the wildlife trade, terrorists could spread a bioterrorist agent or disease, such as anthrax, Ebola or the plague, around the globe in less than 48 hours, without ever having to leave their sitting rooms. As one expert stated when discussing the 2003 monkeypox outbreaks in the USA, 'It was probably easier for a Gambian rat to get into the United States than a Gambian.'[268] Ultimately, such bioterrorist attacks may just end up being the stuff of great thriller novels, but we don't need terrorist groups to threaten us—Mother nature is able to do that all by herself. According to microbiologist Dorothy Crawford, 'microbes are always going to be one step ahead of us. Their generation time is 24 hours, ours is 30 years.'[269]

The wildlife trade will be extremely difficult to eradicate because it is so lucrative and will require cooperation from nations across the world. However, we can take steps to help minimize it. Educating policy makers and the public through media campaigns about the dangers of the trade and the risk of infectious diseases from exotic animals are good steps. We can go further and advocate, in partnership with humane and wildlife protection organizations, for greater restrictions and bans on breeding farms and importation of wildlife. We can, together, also endorse greater enforcement of existing laws and increased fines and penalties. Offering alternative means of sustenance and economic development for impoverished people, such as bushmeat hunters, will help them not only in the immediate future but also further down the road. As wild animal populations become more and more diminished, the livelihood of bushmeat hunters will be threatened.[270] We can help prevent this by fostering the development of more sustainable activities, such as those that take advantage of the rich diversity of their wildlife as a means to entice tourists.[271] Ecotourism has taken off in many parts of the world and, where implemented appropriately to ensure wildlife protection and local participation, is proving rather successful economically.[272] Such activities encourage local populations to help protect their wildlife and provide a more stable source of income.[273]

Ultimately, our greatest power may lie in educating and persuading the public not to buy in to the trade. The trade would not exist if there were not a ready supply of consumers. Educating the public about the health risks and the animal suffering associated with the keeping of exotic animals as pets, entertainment venues using animals, the skin and fur trade, and so on can be a powerful deterrent. Parents will be less likely to take their children to animal circuses and petting zoos if they are aware of the potential health risks posed to their children. We can promote alternative and fantastic sources of entertainment, such as the Cirque de Soleil, which involves no animals in any of its circus shows. Other activities that foster an appreciation of wildlife without causing harm to them include outdoor hiking, animal 'watching', visiting animal sanctuaries and exploring local botanical gardens and their exhibits. The more steps we take to minimize the trade, the greater the chance we have to protect our own health and protect animals from so much harm. As Nathan Wolfe explained, 'Today HIV is so pervasive that it is hard to imagine the world without it. But a global pandemic was not inevitable.'[274] It's too late to prevent HIV, but perhaps we can prevent the next pandemic.

4
Foul Farms: The State of Animal Agriculture

Always remember, a cat looks down on man, a dog looks up to man, but a pig will look man right in the eye and see his equal.
—Winston Churchill

The livestock revolution

As disturbing as the wildlife trade is in fostering the development of new infectious diseases, recent events suggest that the biggest and most imminent threat may lie much closer to home. Between 2007 and 2008, farmers in the Philippines noticed that pigs were falling sick and dying by the hundreds for unknown reasons.[1] A subsequent investigation confirmed the presence of porcine reproductive and respiratory disease syndrome, a serious illness among pigs.[2] But, much to the surprise of the investigators, a subtype of Ebola virus, Ebola Reston, was also discovered circulating in a sample of the pigs. This was the first time Ebola of any strain had been found in these animals. 'We never thought that pigs could be infected,' says Pierre Rollin, an Ebola expert at the Centers for Disease Control and Prevention (CDC).[3] Rollin believes that Ebola Reston is to blame for the pigs' deaths because tissue studies revealed that the virus had pervaded the spleen, similar to its mode of attack in monkeys. Ebola Reston is named after the strain that was discovered in monkeys shipped to laboratories in the USA from the Philippines on several occasions between 1989 and 1996. The first shipment of Ebola virus was discovered after hundreds of monkeys became severely ill or died in a quarantine facility owned by Hazleton Laboratories (now Covance, Inc.) in Reston, Virginia. Because this was the first confirmed instance of

Ebola entering the USA, a panic swept across American health agencies. All remaining monkeys at the facility were euthanized and the building was eventually demolished. Although evidence revealed that humans were also infected, Ebola Reston proved to cause at worst only a mild flu-like illness in humans, unlike all other known strains of the virus. Thus, Ebola Reston, although deadly in monkeys, was deemed only a minor threat to us. However, new fears are rising since Ebola Reston was discovered in pigs.

The industrialization and mass production of animals for food is now among the biggest contributing factors to emerging infectious diseases over the past few decades.[4] Pigs and other animals raised for food are critical sources of zoonotic pathogens that threaten human health and have been directly implicated in the emergence of the H5N1 avian influenza virus, the 2009 H1N1 'swine flu' pandemic virus, the rise in foodborne infectious diseases, and other significant infectious pathogens and diseases. To understand how and why animal agriculture fosters the emergence of new pathogens, it helps to get a glimpse of the experiences of animals raised on modern 'farms'.

Humans are consuming more animals than ever before.[5] Once viewed as a luxury, meat is now becoming a dietary staple for many due to a worldwide growth in urbanized populations and affluence. Today, more than 64 billion animals are raised and killed for food worldwide annually.[6] That means that more than nine farmed animals exist for every human at any one time. China and the USA are among the largest farmed animal producers in the world.[7] In the USA alone, an excess of 9 billion land animals are slaughtered annually for food, approximately 1 million per hour.[8] On average, each American eats the equivalent of 21,000 animals in his/her lifetime.[9] And, global meat production is expected to double by 2020.[10] Consequently, over the last half-century, a dramatic shift has taken place in the animal agriculture industry that may represent the most profound change in the relationship between humans and animals since animals were first domesticated.[11]

In the name of efficiency, the industry has chosen to sacrifice the space and well-being of animals. Traditional farming practices in which animals were permitted to roam outdoors prior to slaughter have largely been replaced by immense, intensive animal operations. These concentrated animal-feeding operations (CAFOs) or factory farms changed the status of animals like nothing had before. The industrialized methods of raising animals for food have spread throughout much of the world.[12] The result is that traditional farms in developing nations are being replaced at a rate of more than 4 percent a year.[13] The independent family farm is rapidly becoming a thing of the past. Today, most animals raised for food are produced by a few immense agribusiness corporations that intensively confine animals by the hundreds or thousands in consolidated operations.[14] This demand-driven transformation of animal agriculture is so dramatic that it has been dubbed the 'Livestock Revolution'.[15]

Factories of misery

By intensively crowding animals into smaller and smaller spaces, agribusiness has greatly increased the production of meat, milk and eggs at a cheap price. However, this increased productivity has come at a high cost to public health and animal welfare. A report by the Pew Commission on Industrial Farm Animal Production, a joint collaboration by the Pew Charitable Trusts and Johns Hopkins Bloomberg School of Public Health, finds that the economies of scale used to justify intensive farming are largely an illusion, perpetuated by the failure to account for associated costs, including environmental degradation and human illness.[16] In these operations, animals are treated as 'production units' and are denied their most basic needs. In many regions of the world, including the USA, many or all animals raised for food are exempt from animal welfare regulations. As a result, the life of the average animal raised for food is categorically miserable from beginning to end. A review of some of the most intensive farming practices will demonstrate how they contribute to the spread of infectious diseases.

'Broiler' chickens raised for meat are confined by the tens of thousands in grower houses, which are typically sunless sheds, barren except for the litter on the floor and rows of feeders.[17] A few operations may house more than 10 million chickens at a time.[18] The sheds are windowless, artificially lit and, because of the dense crowding of animals, they are forced-ventilated to prevent suffocation. Because these birds are selectively bred for rapid growth, up to 30 percent suffer from chronic pain due to bone deformities and joint instability.[19] Leading meat industry consultant and animal scientist Temple Grandin writes: 'today's poultry chicken has been bred to grow so rapidly that its legs can collapse under the weight of its ballooning body . . . I've been to farms where half of the chickens are lame.'[20] In the USA and many other parts of the world, the overwhelming majority of egg-producing hens (an estimated 70–80 percent worldwide) are confined to 'battery' cages, stacked in long rows, several tiers high in warehouses.[21] On average, six birds are jammed into a cage so tight that the average hen lives her entire life in less space than a letter-sized sheet of paper.[22] She cannot even stretch her wings. In addition to the severe restriction in movement, these cages prevent almost all of the hens' normal and critical behaviors, including nesting, perching, foraging and dustbathing.

Regardless of whether or not battery cages are used, most birds raised for food in industrialized countries, including chickens, ducks, geese and turkeys, are confined throughout their lifespan.[23] The crowding and unnatural conditions are so stressful that cannibalistic behavior and injurious feather-pecking are considered a common issue in the poultry business. To prevent the breeding and laying hens from hurting themselves and others—and thus impacting profit—industrial workers amputate the birds' beaks without anesthesia.[24] In contrast, chickens raised for meat are typically not debeaked. This is not due to kindness but because they are slaughtered at only six weeks of age before the previously mentioned injurious behaviors have an opportunity to develop.[25] Male chicks are considered byproducts of the egg-laying industry as they are not specifically bred for meat and are unable to lay eggs.[26] Serving no purpose to the industry, each year billions of them are killed in a cruel manner, such as gassing, crushing or simply piling them into garbage bins to die from dehydration or suffocation.

An estimated half of the world's pigs raised for meat or for breeding are confined to pens or crowded by the hundreds into sheds for the duration of their lives.[27] Breeding pigs are kept constrained for months during pregnancy in gestation crates barely larger than the size of their own bodies, unable to turn, stand or even scratch an itch.[28] In China, the world's leading producer of pork, industrial pig units confine as many as 250,000 pigs in single, six-story concrete buildings.[29] As a routine part of production, pigs are mutilated, by castration and tail amputation, without pain relief.[30]

Although cows have it a little better than chickens, pigs and turkeys, they are now reared en masse in feedlots where they are fattened on unnatural diets and undergo mutilation, such as castration and dehorning, again often without pain relief.[31] Cows in the dairy industry endure intense, prolonged cycles of artificial insemination and mechanized milking for most months of the year in order to produce extraordinarily large quantities of milk.[32] Male calves produced by dairy cows are handed over to the veal industry, known for its extremely inhumane rearing practices.[33] When describing the practices of the animal industry, a Pew Commission member and former assistant US surgeon general, Michael Blackwell, affirmed that 'these animals can't engage in normal behavior at all'.[34]

In the past, farms raising animals for food were highly localized enterprises where animals were bred, raised and slaughtered in the same region or farm.[35] That is a far cry from today's factory farming practices. According to Dr Michael Greger of the Humane Society of the United States, pigs in the USA, for example, 'are frequently born in North Carolina, fattened in the corn belt of Iowa, and then slaughtered in California'.[36] Most pigs and birds raised for meat and eggs never see natural sunlight until the time of slaughter, when they are typically trucked off to slaughter plants, often thousands of miles away. The United Nations' Food and Agriculture Organization (FAO) calls the transport stage 'undoubtedly the most stressful and injurious stage in the chain of operations between farm and slaughterhouse'.[37] It is not uncommon for animals, particularly cows and pigs, who panic and refuse to enter the trucks, to be prodded and beaten into the transport trucks. They are then overcrowded into poorly ventilated trucks or exposed directly to the elements, and suffer from dehydration, starvation, heat exhaustion and freezing temperatures. Once they arrive at the slaughterhouses, their ordeal continues. Birds are typically dumped onto conveyer belts, hung upside down in shackles by their legs and have their throats cut by automated machinery. Because they are exempt from protection in the USA and throughout much of the world, chickens, turkeys, ducks and geese are not typically rendered unconscious before their throats are cut.[38] As many as 8400 birds may run through the slaughter lines per hour, leading to frequent errors in killing; up to 3 percent are still alive when they are placed in scalding water to loosen their feathers prior to plucking. Pigs and cows undergo similar mistreatment as they are shackled upside down and die by bleeding after having their throats cut.

To sum up the realities of animals raised for food, the overwhelmingly majority are housed in extremely filthy, overcrowded conditions without access to fresh air, sunlight or room to move about normally. The animals frequently stand in their own waste for much of their lifetimes and are continuously inhaling and recirculating aerosolized fecal matter, methane and ammonia. It is critical to note that the stress and distress associated with these conditions heightens the animals' vulnerability to disease. A report published by the Pew Commission states:

> Confinement of large populations of animals has several impacts on pathogen risks: first, close contact of large numbers of host animals facilitates the evolution and exchange of viruses, bacteria, and microparasites; second, stresses induced by confinement may increase the likelihood of infection and illness in animal populations;

and third, these large populations produce large amounts of waste, which can exceed traditional methods of management.[39]

How factory farms promote infectious diseases

Almost every aspect of intensive farming contributes to the development of new, re-emerging or more severe infectious diseases. In a paper published in 2007 on the zoonotic potential of factory farms, the authors described 'animal crowding, CAFO hygiene, temperature and ventilation control, and stress all have an impact on growth rate and the ability of animals to resist disease'.[40] Add to this list rampant antibiotic and vaccine use, slaughter practices, stressful transportation and animal mutilation, and manure handling practices and we have a perfect storm for infectious diseases.[41]

It is well established that stress heightens humans' and other animals' vulnerability to infectious diseases.[42] Stress factors are evident across all stages of a farmed animal's life, including the fattening, catching and loading, transport and pre-slaughter handling stages.[43] Physiologic stress has been associated with confinement, stocking density and mutilation procedures, such as castration, debeaking, dehorning and tail docking.[44] Furthermore, stress may not only increase an animal's susceptibility to disease but also facilitate the spread of pathogens.[45] For example, stress has been found to cause damage to animals' intestinal tracts, triggering increased fecal shedding and the spread and emergence of a variety of pathogens, particularly those that cause diarrheal and respiratory diseases.[46] Several studies have revealed that transportation stresses animals, increasing their shedding of pathogens.[47] Transportation has been associated with the shedding of pathogenic bacteria such as enterohemorrhagic *Escherichia coli* (*E. coli*) 0157:H7 and *Salmonella* spp. in fecal matter, resulting in contaminated trailer floors and bedding material, which can cause cross-contamination.[48]

One study by Barham and colleagues assessed the prevalence of *Salmonella* and *E. coli* 0157:H7 within the feces and on the hides of cows before and after trucking. They found that a mere 30–40 minute ride increased *Salmonella* prevalence in feces from 18 to 46 percent and on hides from 6 to 89 percent.[49] The investigators found no increase in *E. coli* prevalence after transportation. However, another study found that out of 286 cows, the percentage with high counts of *E. coli* 0157:H7 present on their hides increased from 9 pre-transport to 70 at the slaughter facility.[50] Similarly, a study of chickens found that, at the time of slaughter, all chickens examined showed gross evidence of fecal contamination on their carcasses.[51] The results of these studies are particularly worrisome since hides and skins are the main source of contamination of processed carcasses. Pathogens present on animals' hides/skins could easily enter the meat supply.[52]

Filthy conditions pervasive in the animal industry also facilitate the transmission of pathogens. The dense concentration of animals in indoor facilities leads to extremely large amounts of aerosolized waste, particularly fecal matter and ammonia. These animals typically spend their days sitting and lying in their own waste. In addition to waste matter, the airborne dust in factory farms has been found to contain viruses, molds, bacteria, bacterial toxins, discharges (vaginal, nasal and respiratory), skin debris, particulate matter and antibiotics.[53] The concentration of these pollutants and fumes is so high that factory farm workers frequently experience a wide range of airway and other diseases, such as asthma, bronchitis, mucus membrane irritation, chronic obstructive pulmonary disease and acute toxicity from high-dose gas exposure.[54] Just a two-hour exposure to the air in these facilities can cause itchy and watery eyes and chest tightness in workers.[55] If only two hours of exposure can cause these symptoms, imagine how a lifetime (albeit artificially shortened) of living in such an environment affects the health of the animals.

Indeed, in a review of the health status of chickens used in meat production, the European Commission's Scientific Committee on Animal Health and Animal Welfare found a clear increase in average mortality over recent decades in chickens due to a combination of diseases, particularly ascites, which is a build-up of fluid in the abdominal cavity that is influenced by air quality.[56] The environment in enclosed battery cage facilities can become so saturated with feces and ammonia that birds develop sores and ammonia burns on their skin in large numbers. The prevalence of footpad dermatitis, a skin disease, in birds increased from 1.4 to 34.5 percent over a 20-year period. Diseases of the skin, respiratory tract, mucous membranes and other organ systems triggered by environmental conditions render chickens more prone to infectious diseases. Gases such as hydrogen sulfide and methane, high air humidity due to poor ventilation, and dust also impact chickens' susceptibility to disease. Similar conditions are leading to sicker and more vulnerable pigs, cows and other farmed animals.[57]

In a review of emerging infectious diseases proliferating as a result of animal agriculture, an international panel of experts stated that a major impact of intensive operations is that they may allow for the rapid selection, amplification and dissemination of pathogens.[58] The panel concluded that 'stated simply: because of the Livestock Revolution, global risks of diseases are increasing'. The panel then went on to say, 'in much of modern society, most people are estranged from agricultural production and have little contact with food animals. Yet, ironically, societal dependency on these animals and vulnerability to them has increased progressively.'

Tainted food

From May to November 2010, approximately 1939 people in the USA fell ill to *Salmonella enteridis* infections, leading to the largest egg recall in US history.[59] *Salmonella* is among the most commonly diagnosed causes of foodborne illness in the USA and is the leading cause of food-related deaths worldwide.[60] In most people, *Salmonella* infections manifest as acute, self-limiting diarrheal illnesses. But in children, the immunocompromised and the elderly, they can lead to more severe consequences, such as arthritic joint inflammation and death.[61] Federal investigators of the 2010 outbreak identified two egg suppliers—Wright County Egg and Hillandale Farms of Iowa, Inc.—as the sources of the contaminated eggs. After the outbreak, the US Food and Drug Administration (FDA) inspected the two farms and detailed unsanitary and inhumane conditions.[62]

The 2010 *Salmonella* outbreak drew national attention to the conditions of industrial egg-laying hen facilities. The focus was on evidence suggesting that the dense confinement of egg-laying hens, particularly in caged systems, substantially increases their risk of contracting *Salmonella*. In the USA, about 97 percent of egg-laying hens are confined to battery cages and just 1 percent are considered 'free range' with access to the outdoors.[63] Almost every scientific study published in the recent past comparing the *Salmonella* risk of different chicken housing systems has found that cage systems pose greater risks than any other housing type.[64]

One of the several factors associated with the increased risk posed by cage systems is the fact that they may be more difficult to clean and disinfect than non-cage systems.[65] Additionally, infestation from flies and other vectors shown to carry *Salmonella* and other pathogens may be more persistent than in non-cage systems, due to manure accumulation under stacked cages and less interference by confined hens.[66] Critics of these studies cite that facilities using cage systems tend to be older than other housing-type facilities and argue that it is the age of the facilities, not the fact that they use cages, that is causing the increased prevalence of *Salmonella* infection in hens.[67] While the age of the facilities may indeed affect *Salmonella* infection rates, there is ample evidence to suggest that a combination of factors, including facility age, flock size and density, sanitation and stress, experienced by the hens play a role in the development, transmission and perpetuation of *Salmonella*. Most importantly, as evidenced by the consistent increased risk of *Salmonella* associated with larger and denser flock sizes in these published studies, it is reasonable to surmise that confining large numbers of animals together poses the greatest public health risk.

The most comprehensive study conducted to date was launched by the European Food Safety Authority (EFSA) to evaluate the public health implications of the European Union's (EU's) move to phase out all battery cages by 2012.[68] An extensive survey was conducted in which more than 30,000 samples were taken from at least 5000 operations across the EU. The presence of *Salmonella* in fecal samples and dust particles was detected in 1486 facilities tested. This translated to an average prevalence of *Salmonella* across hen-laying facilities in the EU of 30 percent (with a prevalence range from 0 percent minimum to 79.5 percent maximum). The study found that without exception, for every *Salmonella* serotype tested and for every housing system analyzed, the risk of *Salmonella* was significantly higher in cage systems than in non-cage systems. Free-range and organic farms had the lowest odds of *Salmonella* contamination of all housing types analyzed. In free-range systems, the odds of contamination with *Salmonella enteritidis*, the most common source of *Salmonella* poisoning in the EU, were 98 percent lower than in cage systems. For *Salmonella typhimurium*, the most common source of *Salmonella* poisoning in the USA,[69] there were 93 percent lower odds of infection in hens raised in organic and free-range systems compared with cage systems.

The EFSA pointed out that the increased risk of *Salmonella* in cage housing may simply be due to the fact that such systems on average house far greater numbers of animals than others, particularly free-range and organic farms. Thus it was difficult to determine whether greater *Salmonella* risk was due to caging animals, flock size or both. Caged birds generally have two to three times less space per bird than cage-free hens. By densely confining animals together, caged systems are able to house much greater flock sizes than other production systems—no doubt causing significant distress to chickens. Research has demonstrated that stress can increase *Salmonella* colonization and spread among chickens.[70] Additionally, cage systems tend to produce larger volumes of fecal dust than others, and this can carry pathogens.

And it's not only the chickens who are impacted: caged hen facilities have been associated with a higher prevalence of respiratory problems in farm workers than non-cage housing systems.[71] *Salmonella* is not just confined to laying hens. In the USA, *Salmonella* species were isolated from about 23 percent of 'broiler' chickens and 9 percent of pigs.[72] Surveys in Canada found *Salmonella* in more than half of all egg-laying flocks and more than two-thirds of 'broiler' flocks sampled.[73]

In addition to *Salmonella*, high stocking densities of farmed animals have been associated with an elevated risk of a number of other pathogens and diseases, many of which can cause human illness, including *E. coli* 0157:H7 in sheep and cows, *Yersinia enterocolitica* in goats, *Salmonella*, *Brucella* and *Cryptosporidium* in cows, and swine flu virus and Aujeszky's disease in pigs.[74] Farmed animals carry many of these pathogens without evidence of illness, enabling the pathogens to circulate among animals undetected unless we are specifically testing for them.

There are several major routes by which humans can be made sick by these circulating pathogens. Humans can become ill by directly contracting pathogens harbored within animal housing units or released into the surrounding environment, through contaminated air, water and waste. Factory farm workers are at particular risk of contracting zoonoses and transmitting the pathogens to their family members and communities. Numerous studies have shown an increased risk of exposure to and infection by both bacterial and viral pathogens among farm workers and their families.[75] These pathogens include swine influenza virus, *E. coli*, hepatitis E, *Yersinia* and *Leptospira*.[76] Lastly, the public can get ill by directly consuming animal products carrying pathogens or by consuming crops and water that have been contaminated by the application of manure used as fertilizer or manure runoffs.

Within factory farms, pathogens are routinely sloughed off from animals' skin and emitted through respiratory, fecal and other excretions into the air and deposited on surfaces. Bioaerosols may be generated as liquid droplets or as dry materials transmitted in the air.[77] A variety of bioaerosol components that can negatively impact human health, including bacteria, viruses, fungi, mycotoxins and endotoxins, have been found in substantial concentrations in factory farms. Once these pathogens and biotoxins are in the facilities, they are released into the external environment through exhaust fans, natural airflow or other means.[78] In a study of a pig confinement operation in the USA, investigators found a marked increase in bacterial concentrations inside the facility and a steady downwind decrease away from the facility in comparison with upwind.[79] Another group of researchers collected 424 air samples from 12 farms in North Carolina over two years and found that the airborne bacterial concentrations on the farms were higher downwind than upwind.[80] Other work has found similar results.[81] Together these studies strongly suggest that factory farms are a noteworthy source of microbial air contamination.

Foodborne illnesses, like so many other infectious diseases, are on the rise. In industrialized countries, foodborne and waterborne infectious illnesses have more than doubled since the 1970s.[82] About one in four Americans gets sick from a foodborne illness each year and more than 1 in 1000 are hospitalized.[83] Worldwide, foodborne microbial diseases kill about 20 million people annually, with animal products topping the list of causes.[84] Several comprehensive reports, including one produced by the Institute of Medicine, have attributed the global rise in the incidence of foodborne diseases to greater consumption of animal products, the intensification of farm operations and rising global temperatures.[85]

Feces in our food

Earlier, it was mentioned that fecal contamination of animal carcasses can cause pathogens to enter our food supply. Fecal contamination of animal products is a common event. Due to their horrific ordeal, animals regularly soil themselves out of fear while they are being slaughtered. The incredibly large numbers of animals slaughtered for food every minute make it impossible for slaughterers or inspectors to examine most carcasses for fecal contamination. And even when obvious fecal contamination is noticed, microscopic fecal contamination is unlikely to be detected. Medical researchers at the University of Minnesota tested more than 1000 food samples from ten retail markets for evidence of *E. coli*.[86] Fecal contamination, as evidenced by the presence of *E. coli*, an intestinal bacterium transmitted via feces, was found in 69 percent of the pork and beef products and a whopping 92 percent of poultry products. Worse yet, more than 80 percent of the *E. coli* recovered was resistant to one or more antibiotics.

Like *E. coli* and *Salmonella*, *Campylobacter* is also carried in the intestinal tracts of animals. The latter is the most commonly identified bacterial cause of diarrheal illness in the world.[87] Foodborne infection in humans usually results in enteritis, but *C. jejuni*, a common *Campylobacter* subspecies carried in the intestinal tracts of chickens, is associated with more severe human autoimmune illnesses, including Reiter's syndrome, an inflammatory arthritis, and Guillain–Barré syndrome (GBS), a disorder that manifests as acute paralysis. It is estimated that one out of every four cases of GBS is initiated by *C. jejuni* infection.[88] Prevalence rates of *Campylobacter* have been reported among poultry flocks to range from 18 percent in Norway to 90 percent in the USA.[89]

There are other ways in which pathogen-carrying animal feces can enter our food (and water) supply besides direct contamination of animal carcasses. In 2006 an early thaw in Brown County, Wisconsin, warmed frozen fields covered with manure. Within days, more than 100 drinking wells were contaminated with *E. coli* and other bacteria, and the water stunk so badly that one resident commented that 'it smells like a barn coming out of the faucet'.[90] Following the contamination, residents suffered from chronic diarrhea, stomach illnesses and severe ear infections. Brown County is one of the USA's largest milk-producing regions, with more than 41,000 dairy cows producing in excess of 260 million gallons of manure every year, which is largely sprayed over grain fields.

Animal manure has historically been regarded as beneficial to the soil, adding rich nutrients and organic matter.[91] Unfortunately, this is no longer the case. Today, due to the immense scale of animal production, we are left with far more manure than we have use for.[92] In the USA, farmed animals produce between 100 and 130 times as much waste as the entire US human population.[93] David Brubaker of Johns Hopkins School of Public Health estimated that a pig farm with 5000 animals produces as much fecal waste as a city with 50,000 people.[94] In contrast to management of human waste, however, there are few regulations for animal waste disposal and no specific requirements for its treatment.[95] The industry's solution to containing all this manure is to hold it in open pits, euphemistically called manure 'lagoons', or to spray the liquid fecal component over crop fields. Seepage from the waste pits and spray areas may contaminate groundwater.[96]

This manure, particularly wet manure, whether lagooned in cesspools or sprayed over fields, provides a nice, cozy environment in which pathogens can flourish and survive for prolonged periods. Manure contains large quantities of *Salmonella*, *E. Coli* and more than 100 other pathogens that can be transmitted to humans and wild animals.[97] Our crops and water sources become contaminated when manure is used as fertilizer, by storm water runoff from sprayed fields, when manure percolates down to the water table and when manure seeps from lagoons.[98]

Cryptosporidium parvum is a protozoan parasite living in the intestinal tracts of mammals and is one of the most frequent causes of water-borne illness. It can survive for more than 250 days in fecal material.[99] A study of pig manure in Canada, the world's leading exporter of pork, found that 26 percent of all liquid manure samples tested positive for *Cryptosporidium*.[100] In Northern Ireland, *Cryptosporidium* was detected in 25 of 56 fecal slurry samples from 33 commercial pig farms.[101] Most pathogens can survive in soil, water and manure in low temperatures for significant periods of time.[102] In general, bacteria can survive for 2–12 months, and viruses for 3–6 months in land-disposed manure.[103] *Salmonella* is one of the most persistent microorganisms in the environment and has been found surviving for 10 months in slurry tanks of cow manure and 3 months in soil after spreading of manure, plowing, and seeding.[104] In California, *Salmonella* was found in the manure of about 70 percent of all egg-laying hen flocks surveyed.[105] *E. coli* 0157:H7, depending on environmental conditions, can survive for more than a year in sheep manure.[106] *Clostridium perfringens*, a rod-shaped bacterium that on rare occurrences causes severe necrosis of the intestines in humans, can survive indefinitely in manure.[107]

Recently, a number of foodborne illnesses that have made thousands of people sick have resulted from the consumption of contaminated fresh produce. The media have covered these large outbreaks extensively, highlighting the poor food-handling practices in many homes, restaurants and food-processing facilities. Public health practitioners have responded to these outbreaks, but most responses have been off the mark. Conspicuously absent from much of the response to these outbreaks is any mention of the fact that, in many cases, industrial animal agriculture is directly at fault.[108] Plants don't have intestines and as such can't produce *E. coli* or *Salmonella*. But they can and do become contaminated with these intestinal pathogens when they are sprayed with contaminated manure or contaminated irrigation water.[109]

In a review of worldwide food- and waterborne outbreaks in the past few decades, investigators Guan and Holley found that when identified, the circumstances that led to the water or produce contamination were frequently either direct contamination by manure or indirect contamination of waterways, drinking wells and municipal wells.[110] For example, humans have been infected with *Listeria monocytogenes* by consuming raw vegetables that had been fertilized with sheep manure.[111] *Listeria monocytogenes* is one of the deadliest bacterial zoonotic pathogens known and is associated with encephalitis, high fatality rates and miscarriages in humans.[112] In the USA, this microorganism is responsible for 28 percent of foodborne illness-related deaths.[113]

Plenty of other foodborne outbreaks have been linked to contaminated manure. The primary source of an *E. coli* outbreak in 2003 from spinach consumption in California, which killed three people and made more than 200 others sick, was traced back to a cattle ranch adjacent to several spinach fields.[114] *Salmonella* has even demonstrated the capacity to invade plant tissue, rendering useless traditional washing and disinfecting methods. In 2001, researchers conducted experiments in which the flowers of tomato plants were inoculated with *Salmonella*, which led to the tomato plant producing tomatoes contaminated with the pathogen.[115] Manure-contaminated irrigation water was likely the primary source of one of the largest recorded outbreaks of *E. coli* O157:H7, affecting more than 700 schoolchildren in Japan after they ate radish sprouts.[116]

Enterohemorrhagic *E. coli* O157:H7, often associated with undercooked beef, has emerged as a significant pathogen since the 1980s.[117] It can cause debilitating disease and death in humans from hemorrhagic diarrhea and kidney failure. It's a natural inhabitant of the intestinal tract of mammals and birds, both of which shed large bacterial numbers in their feces.[118] For humans, the most significant source of *E. coli* O157:H7 is cows. The rise of *E. coli* has been attributed to two primary factors. The first is the intensification of beef production.[119] A number of studies have shown a correlation between the density of cattle and human *E. coli* infections.[120] The second factor is the widespread use of antibiotics in animal agriculture, leading to *E. coli*-resistant bacteria.[121]

Superbugs from the farm

In its 2008 report on animal agriculture, The Pew Commission on Industrial Farm Animal Production wrote,

> antimicrobial resistance is one of the major public health crises of our time. The discovery of [antimicrobials] and their application to clinical medicine are among the triumphs of twentieth century pharmacology and medicine. This triumph has been eroded with the rise and spread of antimicrobial resistance and it has been suggested that we are entering the 'post antibiotic age' of medicine.[122]

A public health crisis is brewing with the epidemic of antibiotic-resistant bacteria and the increasingly diminishing range of drugs available to

combat emerging superbugs (bacteria resistant to multiple antimicrobials). In the mid-1990s, a panel of experts convened for a workshop at
Rockefeller University.[123] In a report summarizing their findings, they
declared that 'after a half-century of virtually complete control over
microbial disease in the developed countries, the 1990s have brought
a worldwide resurgence of bacterial and viral diseases. An important factor in this phenomenon is the acquisition of antibiotic-resistance genes
by virtually all major bacterial pathogens.'

Bacteria are mutating faster than we can produce new antibiotics,
leaving us vulnerable to pathogens against which we have few effective drugs to combat. Former Principal Deputy Commissioner of the
FDA, Joshua Sharfstein, underscored the quagmire we are facing when
he explained that about 90,000 people die every year in the USA from
bacterial infections and about 70 percent of the offending bacteria
display 'resistance to at least one microbial drug'.[124] The Pew Commission looked at a wide range of factors potentially contributing
to antimicrobial resistance and concluded that 'there is considerable
evidence associating antimicrobial use in agriculture with resistant
pathogens in the food supply, on the farm, and in the environment'.[125]

While the EU has banned the use of many medically important antibiotics in farmed animals, much of the rest of the world, including the
USA and China, continues this practice. Because animals in factory
farms are so prone to illness and impaired growth, feeding them antibiotics as growth promoters has become integrally tied up with intensive
farming.[126] The term 'antibiotic growth promoter' is used to describe
any medicine that destroys or inhibits bacteria and is administered at a
low, sub-therapeutic dose.[127] As far back as 1979, the US Congressional
Office of Technology wrote that 'present production is concentrated in
high-volume, crowded stressful environments, made possible in part by
the routine use of antibacterial in feed'.[128] The industry has long relied
on the routine administration of sub-therapeutic antibiotics to control infectious agents that reduce the yield of farmed food animals.[129]
But rather than effectively controlling pathogens, this practice is just
spurring the development of tougher bugs. An editorial in *Scientific
American* acknowledged that 'modern factory farms keep so many animals in such a small space that the animals must be given low doses of
antibiotics to shield them from the fetid conditions. The drug-resistant
bacteria that emerge have now entered our food supply.'[130] Several studies have reported that half of all antibiotics in the USA are fed to
farmed animals.[131] This may be an underestimation, however, as the
Union of Concerned Scientists estimates that the number is closer to

70 percent.[132] Globally, about half of the world's supply of antibiotics is given to farmed animals.[133]

The antibiotics permitted for use in food animal production in the USA and many other countries represent all of the major classes of clinically important antibiotics used to treat human illnesses.[134] Microbes increasingly resistant to antibiotics include some of the very pathogens released from industrial farms: *Salmonella*, *E. coli* and *Campylobacter*. Also included is methicillin-resistant *Staphylococcus aureus* (MRSA), which can cause a severe skin infection called necrotizing fasciitis. Once rare as a cause of necrotizing fasciitis, MRSA is emerging as an increasingly threatening pathogen.[135] The use of sub-therapeutic antibiotics in farmed animals is a highly politicized and charged issue, particularly in the USA. The animal industry and pharmaceutical companies vehemently deny that there is a connection between the use of antibiotics in farmed animals and the rise in microbial resistance to medically relevant antibiotics. Instead they place most of the blame on the overuse of antibiotics in humans, arguing that humans are over- or mis-consuming antibiotics and thus promoting the development of superbugs.[136] Human use of antibiotics does play a role, but there is a strong scientific consensus that routine antibiotic use in farmed animals is also a major cause of increasing antibiotic resistance.

Antibiotic-resistant bacteria can be transferred to humans through three main routes: by consumption of animal products, by direct contact with animals and through the environment. A study in the midwestern USA found that 49 percent of the 299 pigs sampled carried MRSA.[137] About half of the workers sampled also carried MRSA, suggesting transmission between the pigs and farm workers. Since the MRSA isolates found were resistant to tetracycline, the study investigators postulated that MRSA could have been selected by antimicrobial pressure on the farm. Other studies have similarly revealed evidence of antibiotic resistance transfer between animals on factory farms and farm workers.[138] Numerous studies have confirmed the presence of medically relevant antibiotic-resistant bacterial strains in farmed animals and farm workers as well as in the groundwater, soil, crops, air and manure lagoons surrounding factory farms.[139]

Antibiotic-resistant *Enterococci* and *Staphylococci* have been detected in poultry litter and flies collected near chicken houses.[140] Another study conducted by the CDC found multiple classes of antibiotic compounds in pig manure lagoons, and in surface and groundwater collected near chicken and pig facilities, providing evidence that animal waste used as fertilizer for crop fields can be a source of antibiotic residues in the environment.[141] In one of the largest studies performed, investigators compared the drug resistance of *E. coli* from fecal samples obtained from residents of 33 pig farms that routinely used in-feed antimicrobials with ten farms that did not.[142] Resistance was significantly more frequent among residents from farms that used in-feed medication compared with those that did not. In another study, investigators collected air samples from pig factory farms in the USA and found that 98 percent of the bacterial isolates examined were resistant to two or more of the antibiotics that were commonly used as growth promoters in pigs. In contrast, none of the isolates were resistant to vancomycin, which has never been approved for use in animal agriculture in the USA.[143] One study found that bacterial concentrations with multiple antibiotic resistances were found within and downwind of a factory farm even four weeks after sub-therapeutic antibiotics were discontinued.[144] These studies suggest that the antimicrobial drugs to which farmed animals are exposed provide selective pressure that leads to the appearance and persistence of resistant strains.

Antibiotic-resistant bacteria have also been found in animal products, and the transmission of resistant bacteria from farmed animals to humans has been linked to the consumption of these products.[145] A national survey of antimicrobial resistance jointly commissioned by the CDC, FDA and USDA revealed that between 1997 and 2009, *Salmonella* isolated from chicken carcasses at slaughter plants showed increasing resistance to multiple drugs, including ampicillin, cefoxitin and streptomycin.[146] Tetracycline resistance increased from 20.6 to 33.9 percent and ceftriaxone resistance from 0.5 to 12.9 percent. Similar increases in drug resistance were found in cows and pigs. For example, ampicillin-resistant *Salmonella* isolated from cow carcasses increased from 12.5 to 22.5 percent between 1997 and 2009.

Among the strongest evidence linking antibiotic use in farmed animals and the development of pathogen resistance is the temporal evidence suggesting that at least some of the antibiotic resistance that has emerged did so after antibiotics were widely used in animal agriculture. For example, quinolone antibiotics have been used in the USA for decades to treat *Campylobacter* infections in humans, but widespread resistance did not emerge until after fluoroquinolones were approved for use in chicken farms in 1995.[147] In one analysis, the proportion of fluoroquinolone-resistant *C. jejuni* isolates from humans increased from 1.3 percent in 1992 to 10.2 percent in 1998.[148] Ciprofloxacin-resistant *C. jejuni* was isolated from 14 percent of the tested chicken products obtained from retail markets.

The link between quinolone use in animal agriculture and the emergence of quinolone-resistant *Campylobacter* was found to be so strong that the FDA, after a long battle with pharmaceutical manufacturers, finally succeeded in the withdrawal of fluoroquinolone antibiotics for use in poultry farms in 2005.[149] It was the first withdrawal of its kind in the USA. Given the weight of evidence indicating that use of medically important antibiotics in animal agriculture threatens human health, the FDA is trying to take a stronger stance on the practice.[150] Meanwhile, a host of health specialists and consumer groups are advocating the phasing out or banning of the practice.[151] But because intensive farming is so dependent on the use of sub-therapeutic antibiotics to keep animals even marginally 'productive', it is unlikely that the industry's routine use of antibiotics is going to end any time soon. Until the practice of intensive farming itself is drastically altered, we can count on seeing an ever-increasing repertoire of drug-resistant pathogens escaping farms.

Flu farms

Of all the zoonotic pathogens that may escape from the factory farm, influenza A viruses are probably the most worrisome. There are three general categories of influenza viruses (A, B and C) and there are multiple viral subtypes within each category. Human influenza C causes mild disease and has little potential to cause widespread problems. Influenza B circulates only among humans, causes seasonal flu during the winter months and, again, is relatively mild. Influenza A viruses also cause annual flu but can circulate widely among many animal species in addition to humans.[152] Influenza A viruses are classified by subtype based on two proteins on the surface of the virus: the hemagglutinin protein (H) and the neuraminidase protein (N). There are two ways in which influenza viruses can change their proteins. The first is called 'antigenic drift', described as the natural mutation of genetic material over time, which occurs with both influenza A and B viruses. It is associated with seasonal flu epidemics. The second way a virus can change is by 'antigenic shift'. This is a sudden and major change in the virus and occurs only with influenza A viruses. They undergo antigenic shift by rapid mutation of their genes or by reassortment of genes from different influenza A subtypes.[153] The surface proteins of these viruses are continuously under pressure by the hosts' immune systems to reassort and evolve rapidly.[154] Because of their circulation in a wide spectrum of species, there are many subtypes of influenza A viruses, which have a habit of mixing genes and recombining to produce further strains that have never been encountered by the human population. When these antigenic shifts occur, the majority of people have little or no immune protection against these novel strains of the virus. As a result, a pandemic may emerge. Unlike most other zoonotic pathogens, transmission of influenza from person to person occurs swiftly, largely through the respiratory route, leading to influenza being able to infect a large percentage of the world's population in a matter of months. In a public statement, the World Health Organization's (WHO's) Director-General Margaret Chan stated that 'Influenza pandemics must be taken seriously, precisely because of their capacity to spread rapidly to every country in the world.'[155]

The potential for the rapid dissemination of pandemic influenza was recognized in 1918 when a completely novel influenza A virus, first reported in Spain, entered the human population and swept across the globe in record time.[156] The 1918–1919 'Spanish' influenza is still considered the 'mother of all pandemics' and one of the deadliest natural disasters in human history. What made this virus particularly lethal was the fact that it was likely an avian virus that most humans had not encountered before and that caused severe disease.[157] Most often, influenzas are lethal due to the secondary bacterial infections that occur in already weakened individuals. But in the case of the 1918–1919 pandemic, humans were also immunologically naïve and highly susceptible to the virus itself, which caused significant direct organ damage. By the end of the pandemic, one-third of the world's population had been infected and 50–100 million people died, more than the total number killed in all wars of the twentieth century combined.[158] Since the 1918–1919 pandemic, three other influenza pandemics have occurred: in 1957, 1968 and, most recently, in 2009–2010 due to H1N1. While all three caused a significant number of deaths, they fortunately proved far less deadly than the pandemic of 1918–1919. But the fact that pandemics continue to occur reflects the ever-evolving nature of influenza viruses and the question remains: could a pandemic as lethal as that of 1918–1919 happen again?

Bird flu

For centuries, the evolution of the flu virus has remained relatively stable. However, in recent years it has undergone an evolutionary surge, with new variants emerging rapidly. One of the most notable has been the H5N1 highly pathogenic avian influenza (HPAI) virus, commonly referred to as 'bird flu', which was first isolated in a domestic goose from the Guangdong Province of China in 1996.[159] Aquatic birds are believed to be the primordial source of all influenza A viruses.[160] However, people rarely become infected directly from aquatic birds. It is believed that an intermediate host must be involved for the influenza viruses from aquatic birds to be transformed into viruses that can easily infect humans.[161] Chickens and other farmed animals may serve as the intermediate hosts. In fact, avian influenza viruses are partly classified by how severely they affect domestic chickens.[162] Low-pathogenic avian influenza (LPAI) viruses mainly cause respiratory illness and relatively low mortality in chickens. High-pathogenic avian influenza (HPAI) viruses, on the other hand, cause widespread, multi-organ disease and can be highly lethal in chickens.[163] Because these viruses do not normally infect humans, people have developed little to no innate immune protection.

Prior to the appearance of HPAI H5N1, avian influenza was still considered mainly a disease of wild birds with limited significance for humans.[164] The emergence of H5N1 dramatically changed that perspective. The first human outbreak of H5N1 occurred in 1997 in Hong Kong, with 18 cases and 6 deaths.[165] The source of the 1997 outbreak appeared to be the live animal markets where chickens and turkeys—as well as aquatic birds, such as ducks and geese—are sold for human consumption.[166] Since the first human outbreak, H5N1 avian influenza has spread across much of the globe, widely infecting both domestic and wild bird populations.[167] The transport of chickens and other poultry over long distances is blamed for helping spread the infection.[168] Thus far, H5N1 has spread among farmed birds throughout Asia, Africa, the Near and Middle East and parts of Europe.[169] Human H5N1 infections and outbreaks have also spread around the globe, particularly among younger adults and children. Between 1959 and 1997, human cases of avian influenza virus infections were documented on only ten occasions and those cases were relatively mild.[170] Unlike its predecessors, H5N1 is unusually aggressive in humans. As of 22 June 2011 there have been at least 562 confirmed human cases of H5N1 throughout the world.[171] Of those infected, about 60 percent have died.[172]

Luckily for us, although the case-fatality rate of H5N1 is high, the virus has not yet proved highly contagious among humans. Most human cases have occurred through direct contact with infected birds or with their secretions/excretions.[173] Although the virus continues to cause sporadic human infections with limited instances of human-to-human transmission among very close contacts, there has been no sustained human-to-human or community-level transmission identified thus far.[174] However, as the virus spreads among humans and domestic birds, the opportunity for the virus to acquire the necessary characteristics for efficient human-to-human infection through genetic mutation or reassortment escalates.[175] More worrisome is that the intensification of animal agriculture can substantially magnify that opportunity.

H5N1 is not the only highly pathogenic avian influenza virus spreading among farmed birds and/or humans. HPAI viruses have been recognized since the end of the nineteenth century and are known to arise by mutation after an LPAI precursor has been introduced into domestic birds.[176] All avian influenzas start off as mild, low-pathogenic viruses. However, once they enter domestic bird populations, they can rapidly mutate into highly pathogenic viruses. Recent research has shown that even after circulation among domestic birds for very short periods of time, LPAIs can mutate into highly pathogenic viruses.[177]

For more than 100 years, HPAIs have only rarely occurred among domestic birds, but that is now changing.[178] Since 1990, outbreaks of different HPAI viruses of the H5 or H7 subtypes have increased substantially among farmed birds compared with the years prior to 1990.[179] The intensive confinement of birds facilitates both the increasing frequency and the scale of these outbreaks.[180] H7N7 HPAI caused a severe outbreak among farmed birds in the Netherlands in 2003. The virus also infected poultry workers and their families.[181] Although only one person died, there were confirmed cases of human-to-human transmission and fears remain that such a virus could swap genetic material with other influenza viruses or mutate to become even more dangerous.[182] Other notable HPAI epidemics among farmed birds include H7N3 in Canada in 2004, H5N2 in the USA in 2004 and H7N1 in Italy in 1999–2000.[183]

According to experts from the World Organization for Animal Health of the FAO, the poultry industry should have learned two lessons from these prior outbreaks.[184] The first is that if LPAI viruses are allowed to spread, they will eventually mutate into HPAI viruses. The second is that densely confining birds considerably increases their vulnerability to infectious disease. Instead of heeding these experts' advice, however, much of the mitigation attempts are focused on converting small flocks of poultry kept by private households or 'backyard flocks' into commercial confinement operations.[185] No one knows the exact sequence by which H5N1 developed and there has been much speculation that the low-pathogenic precursor to H5N1 from wild birds was introduced to domestic birds through backyard flocks, since many backyard flocks have been affected by H5N1. Backyard flocks (and the live animal markets) can serve as conduits by which avian influenza viruses from waterfowl enter the domestic bird population. But backyard flocks have been used as primary means of raising poultry for centuries without major incident.

Even if backyard flocks played a role in the emergence of H5N1, their contribution to the epidemics of H5N1 have likely been grossly overestimated in comparison with large commercial operations. One major study found that large industrial flocks account for an alarming proportion of HPAI H5N1 outbreaks reported to the World Organization for Animal Health, as compared with backyard flocks.[186] The study authors found that of all the H5N1 outbreaks in domestic birds reported between 2005 and early 2007, 40 percent occurred on farms with 10,000 birds or more even though these large factory farms consisted of only 10 percent of all flocks. In a similar analysis of a 2004 H5N1 epidemic among farmed birds in Thailand, Graham and colleagues found that the odds of H5N1 outbreaks and infections were significantly higher in large-scale commercial bird operations as compared with backyard flocks.[187] Graham et al.'s study confirmed similar findings in studies of HPAI outbreaks in Canada, the Netherlands and Denmark.[188] The authors concluded that 'although the majority of reported HPAI outbreaks in Thailand in 2004 occurred in [backyard flocks], this increased cumulative risk of HPAI in the backyard sector is primarily due to their relatively greater numbers rather than more risky production practices'.[189] In an assessment of the contribution of both backyard flocks and commercial holdings to the 2003 H7N7 epidemic in the Netherlands, the investigators found that backyard flocks were much less susceptible to infection than commercial farms.[190] In an analysis of the 1999–2000 outbreak in Italy, the risk of infection among birds was found to increase significantly with the number of birds confined to a given farm.[191]

The results of these studies should not be surprising given what we know about the conditions on factory farms. Viruses pass readily from animal to animal in these operations and every transmission of the virus to another animal brings us closer to a pandemic. H5N1 demonstrates how a viral challenge emerged from wildlife, adapted to domestic poultry and, after circulating in these populations, acquired limited ability to infect humans.[192] WHO explained,

> highly pathogenic viruses have no natural reservoir. Instead, they emerge by mutation when a virus, carried in its mild form by a wild bird, is introduced to poultry. Once in poultry, the previously stable

virus begins to evolve rapidly, and can mutate, over an unpredictable period of time, into a highly lethal version of the same initially mild strain.[193]

As will be described in more detail later, pathogens can easily enter and exit factory farms. There are multiple routes by which pathogens can enter poultry populations in large commercial farms. For example, after an H5N1 outbreak erupted among chickens in large factory farms in Japan in 2004, flies near the chicken housing facilities were found to carry the same H5N1 strain as that of the chickens.[194] As many as 30,000 flies may enter a chicken facility during a single flock rotation in the summer months. So while open flocks of domesticated birds might serve as a means of introduction of a new LPAI virus, commercial farms can also serve this purpose. Additionally, the chance that a new LPAI virus will become a threat to humans is substantially increased by the presence and conditions of large commercial farms. The greater the number of LPAI viruses circulating in these operations, the greater the odds of their mutating into deadlier viruses with the potential to widely infect humans. As stated by investigators of the H5N1 virus, 'the probability of such a mutation is amplified in the setting of industrial poultry production due to the rapid viral replication that occurs in an environment of thousands of confined, susceptible animals'.[195] The transition of an LPAI to an HPAI virus can result from a single point mutation affecting the H surface protein.[196] 'It can be reasonably assumed' say FAO scientists, 'that the wider the circulation of LPAI in poultry, the higher the chance that mutation to HPAI will occur'.[197] Given sufficiently wide circulation, any avian influenza virus could transform into an HPAI, jump into the human population and cause havoc. WHO warned that 'while the H5N1 is presently the virus of concern, the possibility that other avian influenza viruses, known to infect humans, might cause a pandemic cannot be ruled out'.[198]

Virus mixing vessels

Even as the medical community was bracing itself for the possibility of a pandemic stemming from an avian influenza virus, shock reverberated around the world when scientists discovered that the next pandemic came not from domestic birds but from pigs. The 2009–2010 H1N1 pandemic was the mildest of the four recorded human influenza pandemics. Though there has been much public grumbling concerning the hype surrounding this latest pandemic, there was legitimate reason to take the virus seriously. The 1918–1919 pandemic was also caused by an H1N1 virus and had what was then a unique feature: the near-simultaneous infection of humans and pigs.[199] The virus was found to spread throughout and adapted readily to pig populations.[200] Viral descendants of the 1918–1919 H1N1 virus have been circulating for the past 90 years among pig populations and have been one of the most common causes of respiratory disease among pigs.[201] Despite their harm to pigs, the 1918–1919 descendant viruses were not considered a large threat to humans.

In 1998, however, a completely new virus was detected among pigs in the USA.[202] This was a previously unseen triple-reassortment H3N2 virus, containing genes from avian, pig and human influenza viruses.[203] New virus subtypes have since been discovered in pig populations, many of which have been the result of genetic shift, and evidence suggests that pig populations at some point served as the reservoir from which the 2009 H1N1 virus emerged.[204]

There is no explicit evidence that pigs directly infected humans with the 2009 H1N1 virus. However, the first confirmed case in humans occurred after an outbreak of respiratory illness in La Gloria, Mexico, a town surrounded by factory pig farms.[205] Regardless of the original source or sequence of events, it is clear that pigs can play a major role in the development of new influenza viruses. As previously mentioned, although aquatic birds are considered the primary reservoir for all influenza A viruses, humans are not commonly directly infected by the strains from those animals, as evidenced by the relatively few human avian influenza virus cases reported before H5N1.[206] Pigs, however, are highly susceptible to both avian and human influenza A viruses and are commonly referred to as 'mixing vessels' in whom avian and human viruses co-mingle.[207] In pigs, viruses swap genes and new influenza strains can emerge with the potential to infect humans. The 2009 H1N1 virus contains a combination of genes that was not previously seen in humans or pigs, although it is suspected to have been circulating undetected among pigs for a number of years before it was identified in humans.[208] Six of the eight genes found in the this virus are associated with influenza viruses that regularly cause illness in pigs in North America.[209] The remaining two genes are associated with influenza viruses that were previously only known to be circulating among pigs from Eurasia. Since it contains avian and human genes as well as genes from two different pig populations, the 2009 H1N1 is commonly referred to as a 'quadruple reassortment'.[210]

According to the CDC, the mixing of live pigs from Eurasia and North America through international trade or other means could have created the circumstances necessary for influenza viruses from the two groups of pigs to combine.[211] In fact, a 2009 study reported in *Nature* demonstrated that reassortant influenza viruses with genes from North American and Eurasian pigs were found in samples collected from pigs in Hong Kong as early as 2004.[212] As with avian influenza viruses, there is much reason to believe that the dense confinement of animals is sparking the evolution of these new 'swine' viruses. 'Concerns have been raised', state the authors of one report, 'that rearing many pigs in close quarters can facilitate the introduction and transmission of swine and human influenza viruses in the herd, thereby increasing the chances for the emergence of a [pandemic influenza virus].'[213] WHO and other organizations cite intensive pig farming and other animal factory operations as a significant contributing factor to zoonotic pathogens.[214] As with birds, the crowding of pigs in confined operations increases the transmission of influenza viruses, and occupational exposure to pigs has been shown to increase the risk of swine influenza virus infection in humans.[215] After the 2009 pandemic broke out, *Nature* reported that 'it is now clear that the animal- and public-health communities underestimated the potential for pigs to generate a pandemic virus'.[216]

Just the beginning

Despite the fact that the next pandemic came from an H1N1 virus, we are far from clear of H5N1. Pigs may be the means by which H5N1 gains the ability to widely infect humans.[217] Until recently, there were only sporadic reports of pigs infected with H5N1, but a study published in 2010 confirmed widespread H5N1 infection in these animals.[218] It revealed that between 2005 and 2007, 7.4 percent of 700 pigs tested in Indonesia carried H5N1. The animals may initially have been infected from nearby chicken farms, but there was also evidence of pig-to pig transmission. In the USA, Asia and worldwide, there is an increasing trend to localize pig and chicken farms within the same region.[219] Scientists fear that the close proximity of industrial pig and chicken farms may greatly increase the potential for transmission of pathogens between these two species and increase the odds of an ensuing evolution of a pathogen that could be transmissible to humans. The lack of influenza-like symptoms in pigs carrying H5N1 means that the virus can easily evolve and evade detection in pig populations transported throughout the world. Even more concerning is the fact that one viral isolate from pigs in Indonesia had acquired the ability to recognize a cell receptor present in the noses of both pigs and humans, a change that could allow it to spread easily among people.[220] The study authors surmised that the mutation to recognize the human receptor likely occurred during adaptation of the virus to pigs (from birds). As we saw with the 2009 H1N1, viruses circulating among pigs have less trouble adapting to humans than pure avian influenzas.[221] The discovery of H5N1 in pigs portends that far from being over, our problems with the virus may just be beginning.

This brings us back to Ebola Reston. Ebola is transmitted by contact with body secretions and possibly by the respiratory route, at least in non-human primates (NHPs). Now that Ebola Reston has been discovered circulating among pigs in the Philippines, there is concern that, like influenza viruses, the Ebola virus may mutate into a form that is deadlier to humans.[222] Ebola Reston may also mutate into a form that can be transmitted to humans through the respiratory route, a development that would be perilous. Additionally, because many more people come into contact with or consume pigs as compared with NHPs, there is a much greater chance that humans will become infected with any Ebola strain that emerges from factory farms. Of 141 tested individuals who worked on the pig farms or were exposed to pig products in the Philippines, six showed antibodies to the Ebola Reston virus, confirming that pigs can transmit the virus to humans.[223] No one knows how likely it is that Ebola Reston in pigs will mutate into a form that is lethal to humans but, according to virologist Gary Kobinger, there have been rumors of unusual die-offs in pigs prior to deadly Ebola outbreaks among humans in Africa.[224] Whether there is a connection between the pig die-offs and the Ebola outbreaks in humans is still to be determined. One thing is clear, however: by densely confining animals by the hundreds or thousands, we may unwittingly accelerate a mutation of the Ebola virus. 'When ... viruses get into these confinement facilities, they have continual opportunity to replicate, mutate, reassort, and recombine into novel strains,' says Gregory Gray, Director of the Center for Emerging Infectious Diseases at the University of Iowa College of Public Health:[225]

> The best surrogates we can find in the human population are prisons, military bases, ships or schools. But respiratory viruses can run quickly through these [human] populations and burn out, whereas in CAFOs—which often have continual introduction of [unexposed]

animals—there's a much greater potential for the viruses to spread and become endemic.

When speaking of animal operations, Christopher Olsen, a molecular virologist at the University of Wisconsin, stated, 'now we need to look in our own backyard for where the next pandemic may appear'.[226]

How do we control these pathogens?

With factory farms poised to harbor the next human pandemic, how do we prevent such a catastrophe? In the scientific literature, there are two practices that are frequently offered as solutions to this problem. The first is biocontainment and the second is surveillance. Biocontainment refers to the prevention of the release of a pathogen from the farm.[227] There is a general assumption that large commercial farms are more controlled environments and are better able to contain pathogens than other farm types, such as backyard and free-range operations.[228] Hence, the move to transform backyard flocks into confined housing systems. But converting backyard flocks to confined housing is not likely to result in a major reduction of HPAI risks.

Two recent reports led by public health scientists at Johns Hopkins University examined the ability of factory farms to contain pathogens.[229] They concluded that the very design and operation inherent in large-scale commercial farms compromises biosecurity because of the numerous means by which pathogens can escape and enter a factory farm. These include contaminated manure and water, forced ventilation of airborne dust into the external environment, insect vectors, wild animals and, finally, transfer to farm workers. The necessity for efficient ventilation of densely confined animals greatly impairs attempts at biocontainment. Manure waste piles on farms can attract aquatic birds. Many pathogens have been shown to readily move in and out of commercial operations, even when considerable attempts are made to prevent this.[230] For example, a study in the UK of ten *Campylobacter*-free flocks housed in sanitized facilities demonstrated that despite all biosecurity measures taken, seven of the flocks became colonized with *Campylobacter* by the time of slaughter.[231] Outbreaks of H5N1 have occurred in several commercial chicken and turkey farms with reportedly high biosecurity standards.[232] A single gram of farmed animal feces can contain up to 10 billion infectious virus particles, and just a small amount of infected feces adhering to a worker's boots, for example, may be sufficient to transfer a virus to a susceptible animal population or to the larger community[233] A mathematical model of transmission of influenza viruses among different populations found that when factory farm workers comprised 15–45 percent of the community, human influenza cases in the general community increased by 42–86 percent.[234] Biocontainment, therefore, is unlikely to be an effective preventive measure.

As with the wildlife trade, surveillance of animals is also widely proposed as a means by which we can catch and prevent pathogens from causing epidemics and pandemics, among both animal and human populations. Surveillance, by improved reporting and detection systems, can help us identify some emerging pathogens among farmed animals that may portend trouble. Reporting by farmers in the Philippines enabled scientists to discover Ebola spreading among pig populations. But Ebola was discovered because the number of pigs suddenly becoming sick or dying was sufficient to alert the farmers' attention in the first place. How do we detect pathogens that may be running rampant within factory farms but are not causing any overt signs of illness in animals? Many of the known bacterial and viral pathogens that infect animals remain asymptomatic. H5N1 and 2009 H1N1 caught us by surprise, but they may have been circulating in domestic animals for a long time prior to their emergence as a human threat. Evidence later suggested that the 2009 H1N1 virus had been circulating among farmed animals for years prior to its identification in 2009. How do we monitor for something we don't even know exists? When we don't know what to look for we probably won't know how to look for it.

Each time an infectious disease incident occurs, we learn more about how pathogens transform, disseminate and infect. But we are still a long way from being able to predict which pathogens will become human threats. Numerous pathogens are running amok in factory farms but we do not have the resources or capabilities to detect and monitor every known pathogen (let alone unknown pathogens) and predict which will become threats to humans. Predicting which virus subtype will become the next human pandemic can be extremely difficult.[235] As we have seen, H5N1 is highly lethal in chickens yet has not to this point resulted in a human pandemic. On the other hand, many viral subtypes go undetected or cause only mild disease in animals and yet have the potential to cause great human harm. Predicting which pathogens will pose a human threat is, at best, a guess.[236]

Surveillance can help and is certainly necessary, but it is fraught with difficulties, including scientific and infrastructure limitations. As an example, our public health infrastructure typically relies on the presence of clusters of people falling ill in a specific time period and geographic space to alert us to when a foodborne illness may be occurring at higher than average rates. Most of our food, though, is no longer produced locally. So when eggs produced by a farm in Iowa are contaminated with *Salmonella*, people in each of the states of Wisconsin, California and New York may fall ill, but not necessarily in sufficient numbers to attract the attention of the public health community. For similar reasons, tracing the primary source of the contamination is also proving difficult. Additionally, for surveillance systems to succeed, international cooperation and transparency is also a requirement. However, lack of cooperation has proved to be a major hindrance thus far. These and other difficulties can render pathogen-detection and tracking problematic. The chances are good that we won't know about a new pathogen threat until after it has already entered the human population and caused enough human morbidity to draw attention. By then it may be too late to contain the pathogen.

Killing (culling) populations of infected animals is another popular method used to try to eliminate the spread of pathogens in livestock. Sometimes this works, but many other times it does not. Culling was used in attempts to thwart the spread of H5N1—more than 100 million birds were killed throughout Asia—but it failed. Despite this widespread killing, the next wave of H5N1 re-established itself in the same countries and spread to new ones.[237]

If pathogens are so difficult, or almost impossible, to contain and eliminate, it seems to make more sense to work to prevent them from emerging in the first place. This line of thinking is behind the use of vaccinations and partly behind the use of routine antibiotics in animal agriculture. We know that the widespread use of antibiotics is creating superbugs. Evidence now suggests that vaccinations may be doing the same. Vaccinating farmed animals, while controversial, is on the increase.[238] In order for farm animal vaccine campaigns to work effectively, at least two criteria must be met. First, vaccines must match the circulating viral strain. That's a tall order to meet because viruses are continuously evolving. If, as according to WHO, the vaccines don't match the virus strain sufficiently, the vaccines may actually accelerate the mutation of a virus.[239] Influenza can spread silently among vaccinated birds, both evading detection and coming under novel selection pressures that can produce new strains.[240] Using vaccines against H5N1, for example, could lead to fewer human infections in the short term. But it could also promote further and unpredictable evolution of the virus, thus causing much more human harm in the long run.

The second criterion that a vaccine campaign must meet in order to be effective is that it must prevent transmission of the pathogen to humans. There is no guarantee, even if the vaccine matches the viral strain, that its use will prevent transmission to people. In China, a national H5 poultry vaccination program was implemented in 2005, with documented decreases in outbreaks of H5N1 among farmed birds soon after. But since then, H5N1 cases have continued to occur among both farmed birds and people despite the vaccination program.[241] Egypt similarly instigated a nationwide poultry vaccination program against H5N1, but the virus continues to circulate among backyard and factory farmed birds, and humans are still falling ill.[242]

A report in *New Scientist* noted that 'there has already been speculation that vaccination programs in China may have led to greater genetic diversity in the [H5N1] virus over the past two years, and perhaps even contributed to the current strain'.[243] In 1995 a vaccination program was implemented in Mexico to curtail the spread of a low-pathogenic H5N2 influenza virus that was circulating among farmed birds.[244] Seven years after the implementation of this program, US investigators discovered bird flu viruses circulating among farmed birds that were increasingly different from the vaccine strain. Thus, as the investigators concluded, not only was the vaccine no longer effective but it may have promoted the emergence of new strains.

WHO reports of recent surveillance of farmed birds in China confirm that H5N1 virus subtypes continue to emerge.[245] Health specialists from the University of North Carolina's School of Public Health and the World Organization for Animal Health warn that 'even if an H5N1 vaccine is developed and proven to be effective, there is no guarantee that it will protect against future pandemic strains'.[246] Most pathogens show multiple antigenic variation during the course of infection and, according to Tomley and Shirley, 'the presence of multiple phenotypes or variants of a pathogen is a major hurdle for long-term control by vaccination and it seems that more complete successes such as the eradication of smallpox virus...are aspirational rather than realistic'.[247]

Despite all attempts to contain and eliminate viruses like H5N1, they have not disappeared. Pandemic experts of the Armed Forces Institute of Pathology and the National Institutes of Health in the USA caution,

> Even with modern antiviral and antibacterial drugs, vaccines, and prevention knowledge, the return of a pandemic virus equivalent in pathogenicity to the virus of 1918 would likely kill >100 million people worldwide. A pandemic virus with the (alleged) pathogenic

potential of some recent H5N1 outbreaks could cause substantially more deaths.[248]

Public Health Specialist at the University of Minnesota, Michael Osterholm, reminded us that every day brings the world closer to the next pandemic. He said, 'we don't know if it's going to be H5N1, but there will be another pandemic'.[249]

5

Animal Agriculture: Our Health and Our Environment

He is a heavy eater of beef. Methinks it doth harm to his wit.
—William Shakespeare, Twelfth Night

Climate change and environmental degradation

Even if we could, no matter how unlikely, contain the pathogens running amok among factory farms, we are still faced with a much larger problem. This is because there are just too many animals being produced for food: animals grown for meat and dairy products account for 20 percent of the world's terrestrial animal biomass.[1] To sustain this massive production requires unprecedented quantities of water, energy, land, pesticides and feed crops (crops fed to farmed animals). In exchange for all these depleted resources, we get polluted water, air and land, and perhaps one of the most significant climate transformations in human history.

In September 1999, Hurricane Floyd hit the eastern part of North Carolina, acting as a catalyst for the environmental disaster that followed.[2] As many as 50 animal waste lagoons, some of them several acres in size, filled with floodwaters and overflowed. This manure flowed into the surrounding wetlands and groundwater aquifers, resulting in massive contamination and pollution of drinking water.[3] In addition to multi-antibiotic-resistant fecal bacteria and antibiotics, excess levels of nitrogen nutrients were discovered in the groundwater near some of the farms following the flooding.[4] North Carolina is the second largest pig producer in the USA; pig production in that state alone exploded in the 1990s, growing from 2.6 million pigs in 1988 to almost 10 million today, most of them confined to factory farms.[5] The aftermath of

117

Hurricane Floyd highlighted the disturbing environmental implications of industrial animal agriculture.

In the USA, animal agriculture is responsible for 32 and 33 percent, respectively, of the nitrogen and phosphorus loads found in freshwater sources.[6] The problem with these nutrients is that, like manure, there are too much of them. The application of manure to cropland and leakage from manure lagoons cause nitrogen, phosphorus and other nutrients to run off agricultural land and into waterways. Excess nutrient loading of waterways causes eutrophication, or overfertilization, leading to loss of oxygen, algal blooms and massive die-offs of fish and other animal populations.[7] Most marine algae are harmless. However, the growth of several toxic species of algae is boosted by nutrient supersaturation. These species produce potent neurotoxins that can be transferred through the food web, where they adversely affect and kill fish, birds, marine mammals and humans that either directly come in contact with or consume them.[8] One harmful species, *Pfiesteria*, is believed to be the cause of massive fish kills along the eastern shore of the USA.[9] It produces a potent neurotoxin and an epidermal toxin that have been linked to significant neurological illness and skin lesions in humans.[10] Additionally, nitrogen in manure and liquid waste can contaminate drinking water. These nitrates, which are associated with human health risks, have been identified by the US Environmental Protection Agency (EPA) as the most widespread agricultural contaminant in drinking water wells. Elevated nitrate levels in water can cause severe methemoglobinemia ('blue baby syndrome'), particularly in infants. This is a frequently fatal condition in which the blood has a reduced capacity to carry oxygen.[11]

Excess nitrates and eutrophication render water unfit for drinking. Compounding the eutrophication of our waterways, animal agriculture also consumes 70 percent of the freshwater supply and is among the most damaging industries to the earth's increasingly scarce water resources, contributing significantly to water pollution.[12] According to the EPA, agriculture is 'the leading contributor to identified water quality impairments in the nation's rivers and streams, lakes, ponds, and reservoirs'.[13] In excess of 129,000 miles of streams and rivers more than 3.2 million acres of lakes have been impaired as a result of agriculture, a significant part due to animal waste and factory farms.[14] The primary pollutants associated with animal waste are nutrients (particularly nitrogen and phosphorus), organic matter, solids and odorous/volatile compounds. Animal waste also contains pesticides, hormones and, of course, pathogens and antibacterials. Pollutants in animal waste can impact water supplies through several possible pathways, including surface runoff, erosion, spills, direct discharge to surface waters and leaching into soil and groundwater.[15] Atmospheric transport is another major pathway by which nitrogen and other pollutants are deposited back to the land and waterways.[16] More than 80 percent of ammonia emissions in Europe are generated by animal agriculture.[17] Factory farm waste emits a number of pollutants of concern to human health, including heavy metals, volatile gases, methane, nitrous oxide and hydrogen sulfide.[18]

Factory farm effects on nearby residents

It isn't just those who work in such facilities that suffer from these pollutants. Multiple studies have indicated that in addition to factory farm workers, nearby communities experience a range of health problems associated with its pollutants. A study in North Carolina found increased occurrences of headaches, excessive coughing, runny nose, sore throat, diarrhea and burning eyes among the neighbors of factory farms compared with residents further away from such facilities.[19] Residents near pig farms also experienced reduced quality of life in that they were frequently not able to open their windows or go outside, even in fine weather, because of the poor air quality. Physicians in critical care and occupational medicine studied the impact of a factory farm in Iowa on children.[20] They found that children in an elementary school residing near the farm were five times as likely to be diagnosed with asthma as compared with children from a control school. A study of children attending a middle school where school staff reported factory farm odors both within and outside the school buildings found a higher prevalence of wheezing in the children compared with other middle school children.[21]

The stench from factory farms can be so overwhelming that it greatly affects the quality of life of the neighbors. A study performed by researchers from the Department of Psychiatry at the Duke University Medical Center found that odor produced by such farms had disturbing health consequences.[22] It revealed that people living near industrial pig farms experienced significantly more mood disturbances, such as depression, tension, anger, confusion and more fatigue compared with control subjects. Complaints of unpleasant odors have become increasingly more frequent in communities near factory farms.[23] In a survey of residents near industrial animal farms in Germany, 61 percent complained about unpleasant odors.[24]

Many other studies have found negative health impacts of factory farms on nearby residents.[25] Compounding the immediate health effects that occur in these residents is the fact that these individuals also tend to be disproportionately poorer and have less access to health care compared with those in communities without surrounding farms (most factory farms are in communities with less wealth and political power).[26] This population is especially vulnerable to illness. Their potentially reduced immunity associated with stress from malodor and exposure to the pollutants from factory farms with their more limited health care access makes nearby residents more susceptible to the industry's pathogens.[27] The combination of these factors can facilitate pathogen transmission to larger communities.

Global warming

The environmental impact of factory farms is not confined to their immediate surroundings. The United Nations' (UN's) Food and Agriculture Organization (FAO) published a major report in 2006 entitled *Livestock's Long Shadow*, which detailed the environmental consequences of animal agriculture.[28] It concluded that 'the livestock sector emerges as one of the top two or three most significant contributors to the most serious environmental problems, at every scale from local to global'. According to the report, animal agriculture is responsible for about one-fifth of the world's greenhouse gas emissions as measured in CO_2 equivalents—more than the total amount emitted by the entire transportation sector (which emits 13.5 percent of the CO_2). Animal agriculture accounts for 9 percent of anthropogenic (i.e. produced by human-related activities) CO_2 emissions,[29] but it produces a much larger share of two far more harmful greenhouse gases: methane and nitrous oxide. Some 37, 65 and 64 percent of anthropogenic methane, nitrous oxide and ammonia emissions, respectively, come from ruminant fermentation, manure production, fertilizer use and other factors directly and indirectly associated with animal agriculture. Methane and nitrous oxide have 23 and 296 times, respectively, the global warming potential of CO_2.[30] This means that nitrous oxide is about 300 times as effective in trapping heat in the atmosphere as CO_2 over a 100-year period.[31] Ammonia contributes significantly to acid rain and the acidification of ecosystems by enhancing the harmful effects of sulfur dioxide.[32]

The estimated impact of animal agriculture on climate change based on the above FAO report may be grossly underestimated, however. A more recent report, published in *World Watch Magazine*, suggests that animal agriculture actually accounts for 51 percent of annual world-wide greenhouse gas emissions.[33] The authors of the report offered several reasons for this new estimate, including underestimation by the FAO of the number of farmed animals and their failure to account for the net source of CO_2 produced by:

1. farmed animal respiration;
2. processing of byproducts such as feathers, fur and leather;
3. carbon-intensive medical treatments of millions of cases of zoonotic illnesses secondary to animal agriculture; and
4. deforestation for feed production and grazing, which prevents the reduction in greenhouse gases through photosynthesis.

This new estimate raises important questions about how the global warming impact of animal agriculture is calculated. A study by the National Institute of Livestock and Grassland Science in Japan assessed the effects of beef production on global warming, energy use and water acidification.[34] Its analysis showed that the production of 1 kg (2.2 lbs or about 4–6 cups) of beef released more greenhouse gas emissions than a three-hour car ride and burned enough energy to light a 100 watt bulb for nearly 20 days.

Most of our land is not used directly to feed people but to feed and maintain farmed animals. Animal agriculture takes up a whopping 30 percent of the earth's total land surface, making it the largest single use of land by humans.[35] Some 33 percent of total arable land is used to produce feedcrops.[36] This feedcrop production in turn is responsible for 37 percent of all pesticide use. Animal agriculture expansion is cited as a key factor driving deforestation. For example, 70 percent of previously forested area in the Amazon Basin is now occupied by grazing pastures, with most of the remaining land being used for feedcrops.[37] Through deforestation, land degradation and pollution, animal agriculture is a major driver of biodiversity loss, exacerbating the effects on biodiversity by the wildlife trade and other factors.

The energy input to produce animal products far outweighs the output. For example, producing 1 kg of animal protein requires 100 times as much water as producing 1 kg of grain protein.[38] The US food production system accounts for about 17 percent of the fossil energy used in the country. The average energy input for animal protein production is more than 11 times as much as that for grain protein production. In 2003, scientists at the University of Amsterdam and Loma Linda University estimated that producing 1 lb of beef protein requires more than 10 lbs of plant protein.[39] If these resources could instead be used to produce plant food directly for humans rather than being diverted into intensely energy-consuming animal production, we could, in part, alleviate some growing concerns about a global food shortage.

Because of the climate changes brought about by global warming, food shortages are expected to worsen. Although some regions may see an increase, the net result will be a loss in crop production.[40] That's partly because the regions currently greatly affected by food shortages are also the regions that will be most negatively impacted by extreme weather changes as our planet warms. This temperature increase will, in many cases, adversely impact crop production. Average global surface temperatures have increased by about 0.74 °C (33.3 °F) since the late nineteenth century, according to the US National Oceanic and Atmospheric Administration.[41] Of the 12 years from 1995 to 2006, 11 ranked among the 12 warmest years in the instrumental record of global surface temperature since 1850.[42] The global average sea level has risen since 1961 at an average rate of 1.8 (1.3–2.3) mm/year, and since 1993 at a rate of 3.1 (2.4–3.8) mm/year, with contributions from thermal expansion, melting glaciers and ice caps, and the polar ice sheets.

The Nobel prize-winning UN Inter-governmental Panel on Climate Change (IPCC) wrote a summary report in 2007 of its assessment of climate change. The panel writes, 'Warming of the climate system is unequivocal, as is now evident from observations of increases in global average air and ocean temperatures, widespread melting of snow and ice and rising global average sea level.'[43] The report describes that from 1900 to 2005, precipitation increased significantly in various areas throughout the world, including the eastern parts of North and South America, northern Europe and central Asia. Globally, the areas affected by drought have increased since the 1970s. For the next two decades, a warming of about 0.2 °C (32.36 °F) per decade is projected for a range of future emissions scenarios. Even if the concentrations of all greenhouse gas emissions and aerosols are kept constant at year 2000 levels, a further warming of about 0.1 °C (32.18 °F) per decade is expected. The IPCC projects:[44]

>90 percent probability that heat waves will become more intense and more frequent;

>90 percent probability that heavy precipitation events will become more frequent;

>66 percent probability that tropical cyclones will become more intense;

>66 percent probability that there will be an increase in areas affected by drought;

>66 percent probability that there will be an increase of incidents concerning extremely high sea levels.

Exactly how global warming will be manifested is not clear, but multiple projections suggest that changes in weather patterns will impact food production and water reserves, cause greater malnutrition, diarrheal diseases and death from heat waves, loss of biodiversity, acidification of our oceans, floods and droughts, increased cardiac, kidney and respiratory diseases (asthma, bronchitis) and increased incidence of certain vector-borne diseases.[45] By 2020, between 75 and 250 million people are projected to be exposed to increased water stress due to climate change,[46] and 64 percent of the world's population is expected to live in water-stressed areas by 2050.[47] Although a minority of people are expected to experience health benefits from global warming (mostly due to a reduction in diseases related to cold weather), the global burden of disease and the number of premature deaths are expected to increase.[48] Poorer populations are expected to be most immediately affected by climate change, but more affluent populations will also likely experience substantial increases in, and exacerbation of, health problems, particularly chronic diseases like cardiovascular, cerebrovascular and respiratory illnesses and certain foodborne illnesses and infections, such as Lyme disease and dengue fever.[49]

A major report on the health effects of climate change jointly launched in 2009 by *The Lancet* and University College London, one of the world's leading research-based universities, warned that the greatest impact on global health will come from the indirect effects of climate change on water, food security and extreme climatic events.[50] 'Anthropogenic climate change is now incontrovertible,' says the report. While there is no serious scientific doubt about the reality of climate change and global warming, there is uncertainty about the extent of health and population consequences in the next century.[51] However, the report continues, 'even the most conservative estimates are profoundly disturbing and demand action … less conservative climate change scenarios are so catastrophic that adaptation might be unachievable'.[52] Climate change may be responsible for considerable violence and conflict as a result of greater competition for scarce resources, such as food, water and land.

The public health effects of climate change are already being felt around the world and the IPCC predicts that they will worsen considerably.[53] Already the relatively modest change in climate since the mid-1970s has had a significant public health impact, resulting in an estimated 150,000 deaths and 5,500,000 disability-adjusted life-years in 2000.[54] From drought, to flood, to sea level rise, the tremendous displacement of populations as a result of global warming is poised to create a serious refugee crisis. Some now project that global warming could create as many as 150 million environmental refugees by 2050.[55]

What are our options?

The FAO writes, 'Livestock are one of the most significant contributors to today's most serious environmental problems.'[56] The World Bank has highlighted problems associated with livestock farming that may decrease future animal product consumption. These include environmental degradation, natural resource constraints, increasing energy costs, concerns over endangering global food security, food safety and animal welfare. To satisfy our desire for animal products, we have created a downward spiral. To meet the economic demand, animals are placed into high-density confinement, which makes them ill. We then give the animals antibiotics and vaccines to prevent and combat these diseases, which in turn produces more virulent or drug-resistant pathogens, which threaten our own health. To top this off, factory farms are polluting our water, air and land while consuming massive amounts of energy. How do we stop all of this?

Most strategies to reduce the carbon footprint of animal agriculture focus on technological solutions, such as recycling manure for energy generation or changing animals' feed to less resource-demanding crops. Unfortunately, these ideas fail to address the underlying problems created by animal agriculture. Ultimately, the most effective and comprehensive solution to tackle not only the environmental repercussions but also the infectious diseases and antibacterial resistant pathogens that are emanating from factory farms is to change the demand for animal products that necessitates the existence of these farms. Public health specialists, climate scientists and policy makers are starting to recognize the need to reduce factory farms. After a comprehensive evaluation of the health effects of such farms, the American Public Health Association (APHA), the world's largest association of public health professionals, issued a policy statement calling for a moratorium on building any new confined animal feeding operations.[57] The Pew Charitable Trusts and the

Johns Hopkins Bloomberg School of Public Health have also called for a phase-out of intensive confinement systems.[58] This is a good start but more is needed. According to the FAO report, to avoid an increase in the damage caused by animal agriculture above its present, dangerously high level, the environmental impact per unit of livestock must be more than halved.[59]

Doing so will require radical steps. In order to reduce the number of factory farms, there needs to be a substantial reduction in the demand for animal products. In the *American Journal of Clinical Nutrition* in 2009, public health experts argued that 'considering the surmounting ecologic pressures that a burgeoning human civilization exerts on our planet, there is a need to make hard decisions. Among these hard decisions, many societies and governments in particular, will have to reconsider the increasing demand for an animal-based diet.'[60] Reducing meat consumption would have economic, environmental and public health benefits, including averting the emergence of zoonotic pathogens with pandemic potential.[61]

Our powerful forks

Diets consisting of few or no animal products have consistently proved to be much more sustainable than diets heavy in such products. Just to stabilize greenhouse gas emissions to 2005 levels, a report *The Lancet* advocates the reduction of meat consumption to 90 g per person per day, the average amount of a single hamburger patty.[62] This represents an estimated 40 percent reduction in current meat consumption in developed countries. However, the authors of a more recent analysis of food consumption patterns and climate change argue that 'in the long run, however, achieving further reductions will be necessary.'[63] Taking into account cooking, transportation and production, the study authors found that total greenhouse gas emissions from farm to table are as high as 4.3 and 9.3 kg CO_2 equivalents over a 100-year period for 1 kg of chicken and pork production, respectively. For beef, the emissions are as high as 30 kg CO_2 equivalents. In contrast, the equivalent amount of greenhouse gas emissions for vegetable, fruit and grain production is far less. Producing and cooking 1 kg of soybeans, shipped from overseas, emitted only 0.92 CO_2 equivalents, and carrots as little as 0.42 CO_2 equivalents. Another analysis of food contribution to environmental change found that in comparison with a vegetarian diet, a non-vegetarian diet required 2.9 times more water, 2.5 times more energy, 13 times more fertilizer and 1.4 times more pesticides.[64]

Based on numerous studies demonstrating that plant-based diets are far more sustainable than animal-based ones, health specialists around the world have now called for a radical reduction in meat consumption in order to avert the threats of global warming and environmental degradation.[65] Dr R.K. Pachauri, chair of the IPCC, recently suggested that a reduction in meat consumption would be a practical and helpful way for an individual to contribute to lowering greenhouse gases, which will have the added benefit of helping to reduce the rates of many chronic diseases. A landmark report in *The Lancet* argued that 'governments need to address patterns of food consumption. One starting point is to define and promote a sustainable diet, which could mean reductions of the incidence of heart disease, cancer, diabetes and obesity.'[66] The consensus among many health specialists is that a reduction in animal product consumption would not only help thwart climate change and pollution but also lead to major reductions in chronic diseases most often associated with the consumption of animal products, such as cardiovascular disease, diabetes and many forms of cancer. The FAO agrees that

> a large number of non-communicable diseases among the more wealthy segments of the world's population are associated with high intakes of animal source foods ... while not being addressed by this assessment, it may well be argued that environmental damage by livestock may be significantly reduced by lowering excessive consumption of livestock products among wealthy people.[67]

A 2009 editorial in the *Archives of Internal Medicine* called for a reduction in meat consumption not only to benefit our environment but also to reduce the burden of chronic diseases.[68] As an example, it is estimated that decreasing animal agriculture production by 30 percent would lead to a 15 percent reduction in heart disease in the UK.[69]

Chronic diseases are the largest cause of death in the world.[70] Ischemic heart disease is the number one causes of death worldwide, followed by strokes.[71] Together, they cause more than one-fifth of all deaths. Between 1990 and 2020, mortality from ischemic heart disease in developing countries is expected to increase by 120 percent for women and 137 percent for men.[72] After cardiovascular disease (strokes and heart disease), the top causes of death include chronic lung disease, cancer, diabetes and infectious diseases.[73] A recent article in the *Journal of the American Medical Association* noted that despite the evidence that chronic diseases have surpassed other causes of mortality in developing nations and in the young, 'strong beliefs persist that chronic diseases afflict only the affluent and the elderly'.[74] Overweight and chronic diseases have historically been a concern in more developed countries, but developing countries are rapidly catching up, with obesity surpassing undernutrition as a concern.[75] Globally, more than 1 billion adults are overweight and an estimated 65 percent of US adults are overweight or obese.[76] Of particular concern is the increasing incidence of obesity in children.[77] Being overweight is quickly becoming the norm, and this dangerous trend is associated with significant increases in diabetes, arthritis, asthma, hypertension and hypercholesterolemia.[78] A new study from Harvard University and Massachusetts Institute of Technology projects that by 2050, 42 percent of American adults will be obese—and that's the best-case scenario.[79] The obesity rate might be much higher.

Experts recognize the worldwide transition from a predominately plant-based diet to a diet high in animal products as a notable contributor to the rise in chronic diseases.[80] Animal products are the main source of saturated fats contributing to cardiovascular disease, diabetes and some cancers. They are also the sole source of cholesterol in the diet (plants don't contain cholesterol), which is linked to heart disease and strokes. Eating animal products is associated with numerous cancers. Endometrial cancer risk is associated with increased consumption of total energy, fat and protein from animal sources.[81] Dairy consumption is associated with prostate cancer.[82] The European Prospective Investigation into Cancer and Nutrition study of 142,251 men found that a high intake of dairy calcium and protein increased the risk of prostate cancer.[83] Calcium from non-dairy foods, on the other hand, was not associated with increased cancer risk. In recent studies, breast cancer risk has been associated with higher intake of processed and/or total red meat and with higher intakes of total and saturated fats.[84]

A classic study by Armstrong and Doll published in 1975 revealed a significant association between meat consumption and colon cancer incidence in more than 25 countries.[85] Since that time, the association between colon cancer and meat has been substantially strengthened. Studies in Japan revealed rising incidences of colorectal cancer with greater adoption of Western dietary habits and consumption of meat, milk, eggs, fats and oils.[86] Although confounding factors must also be considered, these and other studies collectively provide strong evidence of the causal link between meat and colorectal cancer.[87] In 2007, the World Cancer Research Fund and the American Institute for Cancer Research panel concluded that there was convincing evidence to recommend limiting red meat intake and following a plant-based diet to reduce the overall risk of cancer.[88]

Substantial evidence suggests that the rise in worldwide obesity and diabetes is associated with increased animal product consumption, in addition to decreased exercise and other factors.[89] Health care costs attributable to meat consumption are considerable. In 1992 the cost was estimated in the US to be between $29 and $61 billion per year.[90] A more recent study found that the economic cost of overweight/obesity is estimated at $300 billion per year in the USA and Canada.[91] According to the American Heart Association, the cost of treating heart disease and stroke in the USA is expected to triple in the next 20 years to $818 billion.[92] Comparative studies reveal that those who follow plant-based diets generally have a lower body mass index and are much less likely to be overweight or obese than those who do not, even across ethnic groups.[93] Additionally, studies show that those who consume plant-based diets benefit from improved health, greater longevity, and a reduced risk and progression of cardiovascular disease, diabetes and many types of cancer.[94]

The American Dietetic Association stated in a 2009 position paper that vegetarian and vegan diets during all stages of the life cycle, including pregnancy and childhood, are 'healthful, nutritionally adequate, and may provide health benefits in the prevention and treatment of certain diseases'.[95] Vegetarians, the position paper concludes, have a lower risk of death from ischemic heart disease, lower cholesterol levels, lower rates of hypertension and type 2 diabetes, lower body mass indexes and lower overall cancer rates. Dr David Katz, Director of the Prevention Research Center at Yale University School of Medicine, argued in favor of plant-based diets, stating that 'the general consensus and weight of evidence tip decisively in favor of diverse health benefits from eating more plant-based diets. Were we to eat more plant foods, and relatively less of all the rest, health, both public and personal, would almost certainly improve.'[96]

According to the WHO, the leading global risks for mortality, affecting all income groups, are high blood pressure (responsible for 13 percent of deaths globally), tobacco use (9 percent), high blood glucose (6 percent), physical inactivity (6 percent) and overweight and obesity (5 percent).[97] These risks are responsible for raising the incidence of chronic diseases such as heart disease, diabetes and cancer. Understanding that animal product consumption contributes to the development of hypertension, overweight and obesity, and high blood glucose, one can estimate that the global burden of chronic diseases attributable to animal products may be as high as 24 percent. In another study, investigators from the Centers for Disease Control and Prevention (CDC) evaluated the top causes of death in the USA and found that while smoking is the leading preventable cause of death, poor diet and inactivity contribute significantly more to the annual death rate than previously believed.[98] The investigators reported that the 'most striking finding was the substantial increase in the number of deaths attributable to poor diet and physical inactivity'. It is clear, the investigators stated, considering trends, that poor diet and inactivity will likely 'overtake tobacco as the leading preventable cause of mortality'. Given that this study was based on 2000 data and that obesity and overweight have since increased, it is highly likely that poor diet and inactivity are now the main contributors to mortality in the USA. Moreover, estimates from both the CDC and WHO reports do not include the global burden of disease secondary to the additional effects of animal product development and consumption: infectious diseases and environmental destruction.

What public health can do

Taken together, these studies reveal that the global burden of disease due to animal agriculture via infectious disease epidemics, environmental destruction, global warming and direct consumption is arguably more than the burden secondary to tobacco use. However, much of the medical field has traditionally been reluctant to directly tackle the health consequences of meat consumption. When this issue has been addressed, focus is usually on mitigation rather than primary prevention, in stark contrast to tobacco control strategies. Mitigation, however, will not have the needed impact as long as the demand for meat continues to rise. The time is ripe for a shift in traditional views of animal product consumption and for new prioritization by health professionals. Public health specialists and health care providers can play a vital role in reversing the trend toward greater animal agriculture production and the affiliated increases in chronic and infectious diseases by promoting healthier, plant-based food choices. We can do so through three main ways: as advocates, as providers and as role models.

As advocates, we can provide medical input into federal policies that affect nutrition and health. The APHA, the American Medical Association and the President's Cancer Panel of the National Cancer Institute have highlighted the importance of federal food policies in affecting the obesity epidemic, as well as the importance of public health leadership in federal nutrition policy reform.[99] As an example, currently

US agricultural policy disparately promotes animal products, in contradiction to the US Dietary Guideline's emphasis on plant-based foods. Some 73 percent of the more than $60 billion in federal commodities payments for domestic food consumption between 1995 and 2005 supported the production of meat, eggs and dairy, either directly or indirectly through feedcrop supports.[100] Less than 0.5 percent of federal subsidies were allocated to fruit and vegetables.

Studies have demonstrated that changes in agricultural subsidy policy can abate rising chronic disease rates. In Poland, the withdrawal of large animal product subsidies led to decreased saturated fat intake and increased fruit and vegetable intake.[101] Between 1986 and 1994, there was a 23 percent decrease in animal fat availability and a 48 percent increase in vegetable fat availability. Between 1991 and 1994, the importation of certain fruits doubled. Poland's change in subsidy policy was followed by a subsequent decrease in ischemic heart disease mortality. After long periods of increases, mortality from heart disease and stroke decreased by 25 and 10 percent, respectively, between 1991 and 1994 among those aged 45–64.

A greater public health voice in environmental policy, designed to limit further animal agriculture development and/or place restrictions on harmful factory farm practices, would also be helpful. For example, Andy Burnham, former Health Secretary in the UK, calls for a 30 percent reduction in the number of farmed animals bred.[102] Public health coordination with animal humane organizations and environmental groups can provide a great boost to efforts to ban the most egregious of animal housing and rearing practices, such as battery cages, and restrict antibiotic use, both of which increase our risk of infectious diseases. In addition to requiring nutrition labeling on food products, 'carbon-footprint' labeling is one potential way to educate consumers about how their food choices affect climate change.

As providers of health care, practitioners can routinely counsel their patients about nutrition. Several studies have demonstrated that when health care providers advise their patients about nutrition, the incidence of chronic diseases may decline.[103] Despite the potential of medical counseling to improve dietary practices, many primary care physicians never include nutrition or dietary counseling in their patient visits, or include only perfunctory counseling. Medical societies can facilitate the incorporation of routine nutrition counseling by providing guidelines and educating health care providers.

Health professionals can set an example by consuming fewer animal products at home and at work, and by demanding healthier plant-based options in hospitals, in cafeterias, in doctors' and nurses' lounges, and at professional conferences and meetings. The APHA and the Johns Hopkins Center for a Livable Future offer prime examples of how meatless meals can be promoted.[104] Through the work of the Health Care Without Harm coalition, over 122 hospitals in the USA have signed a pledge to offer healthier food items to visitors, patients and staff.[105] Meatless Mondays is now an international campaign to promote healthier and more sustainable food options, and it has encouraged many to eliminate animal-based foods from their diets altogether. Increasingly advocated by celebrities, athletes, chefs, environmentalists, business icons and even former US president Bill Clinton, eschewing animal products is one of the biggest dietary and consumer trends worldwide.[106]

A reduction in or elimination of animal product consumption might seem an extreme step to some, but it is undeniably the easiest thing we can do and the most comprehensive solution we have to so many of the health, environmental, humanitarian and animal welfare problems we face. There is no other option which will single-handedly help thwart food-shortage crises in parts of the world, increase longevity and dramatically decrease rates of chronic diseases, keep us slimmer, prevent epidemics and pandemics, decrease tremendous suffering in animals and maybe even keep the planet from overheating. Bill Jeffrey of the Center for Science in the Public Interest astutely pointed out that despite knowing how much prevention power we have by altering our eating habits, 'we don't have public health authorities that are prepared to press government down in that direction and take direction on their own.'[107] Are we, as the voice of public health, ready to take direction on our own? Given how much we can accomplish by simply directing our forks toward one plate of food over another, it would be a calamitous shame not to pursue a great reduction in, if not complete elimination of, animal consumption as a public health priority.

6

The Costs of Animal Experiments

> The important thing is not to stop questioning.
>
> —Albert Einstein

A major glitch in drug development

Despite our tremendous prevention power, the fact remains that drugs are important tools in the arsenal of modern medical science. To produce new drugs, we need research. This involves applied research, that is, research directly intended to produce a new treatment. Basic, or more exploratory, research is also utilized to help direct applied research. To approve a drug for the market, regulatory requirements usually dictate at least two major stages of safety and efficacy testing. The preclinical stage includes the use of in vitro and/or animal experiments to assess whether a drug is a viable candidate for further clinical investigation based on safety and efficacy evaluations. The clinical stage is broken down into three phases. Phase 1 typically involves a small group of healthy human volunteers to test the safety of a compound. Phases 2 and 3 usually include larger groups of volunteers in controlled clinical trials to test for both safety and efficacy of the potential treatment against the targeted disease or condition. Post-marketing studies are also often required to monitor the safety of a product once in use.

The USA leads the world in the amount of resources directed toward biomedical research.[1] It spends an estimated $100–$120 billion on research annually. Pharmaceutical industries are the largest contributors to biomedical research spending with the publicly funded granting agency, the US National Institutes of Health (NIH), being the second largest contributor, funding approximately $31 billion per year of research.[2] Money spent on biomedical research is growing so fast that it outpaces growth of the gross domestic product in the USA.[3]

Despite the impressive amount of money being spent on biomedical research, the USA lags behind 41 countries in life expectancy.[4] Clearly something is amiss. Much of the reason why the USA lags in longevity is the relatively low priority it gives to disease prevention in comparison with many other developed nations. However, a major glitch in the drug development world has also been increasingly noted. In 2006 an article in the *Journal of the American Medical Association* (*JAMA*) reported, 'While investment in basic research in the United States doubled from 1993 to 2003, the number of therapeutics entering the clinic has actually declined.'[5] New compounds entering phase 1 trials today have about an 8 percent chance of reaching the market.[6] Many drug candidates that enter later phases of the drug development process are also falling by the wayside. A recent analysis revealed that in phase 3 trials the failure rate is now 50 percent.[7] Overall, 92 percent of drugs that pass preclinical tests fail to make it to the market because they are proved to be ineffective and/or unsafe in people.

From 1996 to 1999, 157 new drugs were approved in the USA. A decade later, from 2006 to 2009, only 74 new drugs were approved. Of all these approved drugs, not one of them, according to a recent report, was a cure or a meaningful novel treatment for a host of serious diseases.[8] This has led many to voice concerns about the stagnation in production of useful treatments—and this concern is nothing new.[9] Memorial Sloan-Kettering colon cancer specialist Leonard Saltz lamented the lack of cancer treatment breakthroughs when he said that despite all the hype and excitement about pricey new cancer drugs, by far the most important colon cancer drug remains a 50-year-old chemotherapeutic drug called 5-FU.[10]

There are several potential reasons offered for the reduced number of treatment approvals, including higher regulatory hurdles, longer and more expensive clinical trials and less flexibility in pricing.[11] Perhaps the most salient reason, however, is that noted by the Institute of Medicine (IOM). In June 2000 the IOM conducted a clinical research round table to discuss the state of medical research.[12] The IOM pointed out a 'disconnection between the promise of basic science and the delivery of better health'. In essence, basic biomedical research is generally not efficiently leading to therapies, despite our significant investment of money, time and other resources. It was reported in *JAMA* that because of a doubling of the NIH's budget in recent years as well as major new advances in basic research, many have made the assumption that progress was being made that would result in improved human health.[13] The report then goes on to state that this assumption has been an illusion. Both John Ioannidis from Tufts University School of Medicine in Boston and an article in *Drug Discovery Today* echoed this sentiment and commented that while basic sciences are believed to have made major progress, this has not resulted in the same level of progress in understanding the clinical basis of diseases or in developing novel effective treatments.[14]

In summary, while the pace of basic biomedical research has been rapid, it has not translated effectively to new therapies that have a measurable impact on our health.[15] Something is not working. Why is our tremendous investment in biomedical research not returning on its promise? Two investigators looked at the overall lack of successful development of drugs to treat a host of central nervous system (CNS) disorders.[16] In recent years, only 9 percent of CNS compounds that enter phase 1 clinical trials survive launch. The investigators concluded that one of the main reasons for this high failure rate is that animal models are a far from perfect predictor of drug efficacy in humans:

> The increasingly high failure rates of CNS compounds in human trials has demonstrated that this success in animal models is no guarantee. No animal model is a perfect mimic of human disease. Animal models can serve as models of disease mechanisms, but not of the disease itself . . . Failure rates in clinical development attest to the disparity.

Over the years, much of biomedical research has moved away from more directly studying human physiological mechanisms and diseases and instead has focused on creating and studying models of diseases and mechanisms in animals. There is now a growing recognition that there is an incongruity between understanding mechanisms in animals and understanding an actual human disease.[17] Investigators from the Department of Clinical Neurosciences at the University of Edinburgh noted that while the mechanisms of stroke in animals are well understood, this has not translated to positive results in humans.[18] More than 350 interventions have published efficacy in animal stroke models, of which around 100 have been tested and proven ineffective in human stroke studies.[19] Thus, as illustrated by this one example, understanding disease mechanisms in animals, whether by creating animal models of diseases or though basic physiological research, is not successfully leading to new therapies. 'The failure of neuroprotective drugs in clinical trials,' commented one publication on stroke studies, 'represents a major challenge to the doctrine that animals provide a scientifically valid model for human stroke.'[20]

The lack of sufficient success in utilizing animal experiments to yield new therapies is a fact not just in the field of stroke or basic research but also in applied research. Researchers from the Animal Bioscience and Biotechnology Lab from the US Department of Agriculture (USDA) provided a frank appraisal of the usefulness of animal experiments in predicting human outcomes and found that, on average, 'the extrapolated results from studies using tens of millions of animals fail to accurately predict human responses'.[21] Even the use of multiple species of animals frequently fails to predict efficacy in human trials.[22] In 2002 several leaders in the biotechnology and pharmaceutical industries published a paper outlining what they saw as the major problems underlying the drug development process. They concluded that the poor predictability of animal experiments is one of the major challenges facing the drug discovery community.[23] Based on these and a host of other reports, many in the health community are arriving at a harsh realization—we are failing to effectively discover new therapies in large part because of our focus on animal experimentation in biomedical research.[24]

Over the past few decades, evidence-based medicine has become the mantra of sound, scientifically based medical research and practice. We rely on evidence-based medicine in virtually every facet of health research and practice save one—the use of animal experimentation to inform human health. Animal experimentation has not been subjected to the kind of scrutiny it requires.[25] It is most often viewed as the default and 'gold standard' method of testing, yet it doesn't, with few exceptions later described, receive the critical examination needed to determine its relevance to human health.[26] As a result, there is a dearth of published, peer-reviewed evidence to support the usefulness of animal experimentation.[27] The lack of critical studies examining the relevance of animal experiments was reflected in a recent report from the Nuffield Council on Bioethics.[28] Instead of critical examination, anecdotal evidence or unsupported claims, which are inadequate forms of evidence for a scientific discipline, are substituted as justification for animal experiments.[29]

When animal model validity is discussed, it is usually in terms of the similarities between the model and the human condition it is intended to mimic. However, very infrequently is any formal validation of such models applied.[30] A review of the published literature revealed that even in cases when an animal model(s) is alleged to replicate a human condition, there were very few studies that formally evaluated the ability of these models to reproduce the human diseases in question.[31] In an article published in *Slate* magazine in 2006, entitled 'Of Mice and Men:

The Problems with Animal Testing', reporter Arthur Allen expressed this concern about the reliability of animal experiments to predict harmful adverse effects of drugs in humans: 'Surprisingly, although it is central to the legitimacy of animal testing, only a dozen or so scholars over the past 3 years have explored this question. The results, such as they are, have been somewhat discouraging.'[32] When we actually scrutinize animal experiments, we discover that they are far from the panacea we believe them to be. As a result, a growing number of scientists are questioning the relevance of animal experiments as they relate to human disease and their ability to lead us down the right path toward effective treatments to improve human health. Scientists have also highlighted several notable shortcomings with, and obstacles to, the use of animal experimentation to inform human health. These obstacles include the effect of the laboratory environment and other variables on animal physiology, and thus study outcomes; disparities between animal models of disease and human diseases; and species differences in physiology.

The many influences on animal experimental results

In 1995 Superman actor Christopher Reeve became quadriplegic after being thrown from a horse. He turned his tragedy into advocacy and galvanized the public and scientific community to invest in spinal cord injury research. Unfortunately, there was no substantial return on that investment during Reeve's lifetime. In 2004, following his death, *New Scientist* reported that in 2000,

> [Reeve] pointed out that it was an exciting time for the field— a time when he heard that researchers could cure a rat with a spinal cord injury. Sometimes, the actor said, he wished he were a rat... Following Reeves death this week, his rebuke seems as fitting as ever. While basic neuroscience research is booming, there are precious few treatments—let alone cures—for people with diseases of the brain and nervous system.[33]

Reeve's comment about curing spinal cord injury in rats was not far from the truth. Multiple neuroprotective agents have been successful in treating spinal cord injuries induced in animals in the laboratory. Yet they have all produced extremely disappointing results when tried in humans.[34] The clinical usefulness of one treatment being used in humans, methylprednisolone (MP), is hotly debated. The jury is still out as to whether or not it causes any meaningful reduction in damage following spinal cord injuries in humans. In order to assess whether experiments on animals provided any clarity to the issue, several colleagues and I conducted a systematic review of all published animal experimental studies using MP to treat spinal cord injury and broke the results down by species.[35] The review found results differed between species and among strains within a species.

The question that then followed was: do we pool results from all tested species and experiments, or do we put our faith in the results from certain species and experiments we believe to be most predictive of human responses? If we choose the former, our answer on the usefulness of MP may depend on whether most of the animal experiments involved rats, which showed mostly negative results (i.e. the treatment was not effective), or cats and dogs, which showed mostly positive results (i.e. the treatment was effective). If instead we decide to put our faith in test results from species and experiments we believe to be most predictive of human responses, how do we know which species to choose, and which set of results do we decide are most applicable? The set of experiments conducted in rats or the ones using dogs and cats? But it doesn't stop there—do we trust the results from a certain *strain* of rat and not *another strain*? These are not questions to be taken lightly—answering them is critical to determining which animal experiments best predict human results. Unfortunately, situations in which we know in advance which species or which animal model is most predictive of human outcomes are exceedingly rare, if they exist at all.

My colleagues and I then conducted an investigation to explore the potential reasons for the wide variety of results between and among species and found that many factors in the experimental protocol affect study outcomes.[36] These include how animals are handled, housed, fed and tested, and what type of anesthesia is used during injury induction. For example, cage conditions were found to affect recovery from spinal cord injury in animals. Environmental conditions can influence neurogenesis, gene expression, signaling between nerves and behavioral responses, all of which can significantly impact the results of a study. Even more surprising was that the type of flooring on which an animal is tested or whether or not there are other animals in view of the tested animal can affect whether a drug shows a benefit or not. These unintended influences go beyond studies in spinal cord injury. In a study of a genetic mutation that causes defects in the aorta, the type of environment in which mice were housed affected whether they developed the defects.[37] Another study showed that even modest differences in housing for just one month led to structural and biochemical differences in the brains of two groups of marmoset primates.[38]

Stress, housing environment and diet can all affect study outcomes.[39] These conditions and factors can affect study outcomes in ways that experimenters may not understand, be aware of or be able to control. Even routine laboratory procedures and conditions, such as blood collection, noise produced in the laboratory, cage components and handling by experimenters, can lead to pronounced and/or prolonged changes in genetic expression and stress-related physiologic markers.[40] Ventilation and ambient noise can produce stress and affect an animal's physiology. For example, noise levels of 90 dB (about the sound of a kitchen blender), which is not infrequent in the laboratory setting, have been found to increase heart rate and blood pressure, and to damage small blood vessels.[41] Even the time of day when animals are tested can give different results. In a study of mice, motor deficits (weakness) were evident only at one time at night.[42] Experiments performed on rats in the spring can generate very different results from those performed in the late fall.[43] Tests can be affected by many additional laboratory factors, including environmental humidity, cage density and within-cage order of testing.[44]

The Scientist acknowledges that the laboratory environment can influence the results of an experiment.[45] It reported that many of the underlying limitations associated with animal experiments involve the inherent nature of animal testing. The laboratory environment can have a significant effect on test results, as stress is a common factor in an animal's life in the laboratory. Jeffrey Mogil, a psychology researcher, also demonstrates that the very presence of a researcher alters behavior in mice, which could have an impact on study results.[46] Every procedure and every environmental element in a laboratory setting can, and likely does, influence what results a study produces. Unlike with humans in clinical trials, we can't tell animals to ignore one factor or another—we have little control over their reactions to different procedures and situations. Additionally, and most importantly, animals in laboratories have little to no control over their environments, to which they are exposed, on average, for the duration of their lives. Animals' lifelong exposure to the laboratory setting increases the likelihood that such settings will substantially affect their physiology in unpredictable ways.

For the above reasons, many have called for the standardization of laboratory settings and procedures.[47] The problem as it applies to animal testing is that there are simply too many variables to achieve true standardization. Many of these—most notably those that produce

significant stress, such as catching, restraining and blood collection—
are unavoidable.[48] A study published in *Science* found that despite
all attempts to standardize the environment across three laboratories,
there were systematic differences in test results.[49] What's more, different mouse strains varied markedly in all behavioral tests, and for some
tests the magnitude of genetic differences depended upon the specific
testing laboratory. Controlling how animals react, whether physiolog-
ically or behaviorally, to the procedures and settings in laboratories is
unattainable in any practical sense.

Ultimately, the attempt to standardize laboratory settings and procedures fails to address the fundamental issue, which is not to improve
comparison between labs but to improve the predictive value of experiments to the human condition. As increasing numbers of studies
reveal discrepancies between animal experimental and clinical trial
results, many scientists are requesting that more rigorous methodologies and practices (in addition to standardized environmental settings)
be applied to animal experiments in an effort to reduce the discrepancies. These more rigorous practices would include assurances of adequate
study power, randomization and blinding, and minimization of bias
in publications.[50] Yet, although a step in the right direction, the call
for improved methodologies minimizes another, more important and
unmodifiable limitation of animal experiments—the animals themselves. In the review of spinal cord experiments using MP previously
described, subgroup assessment was conducted on only the animal
experiments that were of the best quality (e.g. those that included
blinding and randomization, and reporting of housing and handling
procedures) and used the same dosing and regimen of MP treatment.[51]
Despite this, study results still varied considerably, indicating that no
matter how methodologically superior and standardized the experiments were, factors inherent in the use of animals accounted for some
of the major differences in results.

Returning to stroke for a moment, many questions have been raised
as to why more than 100 potential therapies failed to translate successfully from animal experiments to human trials. Acknowledging the
failure of finding new, effective stroke treatments despite so many successes in animals, a set of guidelines was implemented by a stroke round
table in 1999 to standardize and improve the applicability of stroke
experiments in animals to humans.[52] One of the most promising stroke
treatments later to emerge was NXY-059, which proved effective in animal experiments. In 2006, at the Joint World Congress for Stroke held
in Cape Town, South Africa, news spread quickly that NXY-059 fell

victim to the same fate as so many prior drugs: it failed in clinical trials. It failed despite the fact that the set of animal experiments on this drug followed the guidelines set forth by the round table and was considered the poster child for the new experimental standards.[53] 'There's no doubt about the absence of an effect of [NYX-059], and that called into question the many other studies in stroke, and how good are the animal models?' said one of the clinical consultants to the trial.[54] Despite earnest attempts, standardization and improvement of animal experimental methodologies hasn't eliminated the substantial discrepancies between animal experiments and human results.

Incongruencies between animal models and human disease

In addition to the unpredictable influences of laboratory environments on animal experimental results, the lack of sufficient congruency between animal models and human disease is another frequent and significant obstacle. When we try to create stroke in animals, for example, we artificially create a disease that occurs naturally in people. The inability to reproduce the complexity of human diseases in animals is a crucial hindrance to their use.[55] Even if design and conduct of an animal experiment are sound and standardized, the translation of its results to the clinic may fail because of disparities between the animal experiments and the clinical trials.[56] In stroke research, these disparities include the presence of pre-existing diseases and conditions in humans, but not in animals, that affect the development of stroke, such as diabetes and atherosclerosis; use of additional medications to treat these risk factors in humans; and nuances in the pathology of the human disease that are absent or different in animal models. Other disparities cited include the use of young and male animals for diseases of the elderly or women.

As a result of the recognition of these discrepancies, several publications argue for the need to use animals who are matched in relative age and gender to the target humans, who are given the same medications as those given to human patients and who have also been altered to manifest the pre-existing conditions (and co-morbidities) that occur naturally in humans.[57] If we try to reproduce the pre-existing conditions in animals, we still face challenges regarding the inability to replicate their complexity. For example, stroke and heart disease are frequently a result of atherosclerosis. Most animals in laboratories don't naturally develop significant atherosclerosis, which is characterized by a narrowing of blood vessels by plaque build-up. In order to reproduce the effects of atherosclerosis in animals, researchers ubiquitously clamp their blood vessels. Simply clamping blood vessels, however, does not replicate the elaborate pathology of atherosclerosis and the causes behind it. In attempting to reproduce the complexity of human diseases in animals, we need to reproduce the complex physiology of the predisposing diseases and conditions, which also proves difficult to accomplish. Thus we end up continuously chasing our own tails. Each time an animal model fails to successfully translate to humans, no shortage of reasons is proffered to explain what went wrong—poor methodology, lack of relevant pre-existing conditions and medications, wrong gender or age, and so on. Recognition of each potential difference between the animal model and the human disease creates a renewed effort to eliminate these differences. What is too often ignored is that these models are intrinsically lacking relevancy to the human diseases they are intended to reproduce.

As early as 1990, major discrepancies between animal models of stroke and stroke in humans were noted.[58] Several neuroscientists asserted that animal stroke models are severely simplistic in comparison with the human disease and labeled stroke animal models a failed paradigm, arguing instead for human-based research.[59] Given the continued failure of animal stroke experiments to unravel new, effective human treatments, and despite all attempts to improve their human relevancy, the sentiment expressed in 1990 remains salient today. Naturally occurring diseases are far more complex than what is produced when we alter a few mechanisms in an animal. Even with diseases for which there is great mechanistic understanding, there can still remain significant disparities between the animal models used and the human diseases being targeted for treatment.[60]

Consider animal models of Alzheimer's disease (AD). In humans, AD is characterized pathologically by the presence of several key features in the brain. A truly predictive animal model must reproduce the origins or etiology, the physiologic basis, the pathology and the symptoms or signs of the disease.[61] Experimenters have altered genes in mice to create models of AD. But herein lies the problem: each mouse model is different and no single mouse model shows all the pathologic features of AD.[62] Instead, each model displays bits and pieces of Alzheimer's and many display features not present in human AD. Consequently, these models often give conflicting results because they differ in regard to the signs that manifest and the causes behind these signs.

Substantial effort has been made to improve the relevancy of AD mouse models. Despite these attempts, these new mouse models still fail to appropriately mimic what occurs in humans.[63] The lack of congruency between mouse models and the human disease may cause potential drugs to seem to be ineffective, while it's actually the mouse model that is to blame.[64] One of the key messages of the 2007 Inaugural Alzheimer's Drug Discovery Foundation Meeting was that the patient is currently the only true model of AD.[65] Existing animal models replicate various aspects of the disease but do not fully mimic the human condition, resulting in a low predictive value. The conference further concluded that using models with low predictive value provides little understanding of the pathophysiology (the physiology and functional changes) of a disease. One investigator commented that 'in reality, disease models usually model only certain aspects of clinical symptomatology, and because only rarely is the etiology of diseases well understood, the induction of the disease state in the model can differ from the clinical condition'.[66] In other words, because we rarely fully understand how and why a disease occurs in humans, when we try to replicate that disease in animals we are usually falling well short of the mark. We take a few observations from humans then try to recreate those observations in animals, and we end up relying on the animal models in place of understanding the full disease in humans. This illustrates a fundamental flaw in our use of animal experiments: we are usually studying models that are at best very incomplete or at worst contrary to the human disease. Either way, the models are incorrect.

David F Horrobin, an influential figure in drug development, commented on the obstacles the pharmaceutical industry faces in delivering new therapies and criticized assumptions made about the congruence of animal models of disease to human diseases.[67] For an animal model of disease to be congruent with the human disease, he argues, three conditions must be met:

1. we must fully understand the animal model;
2. we must fully understand the human disease; and
3. we must have examined the two cases and found them to be substantially congruent in all important respects.

Horrobin contends that these three conditions have not been fulfilled for any human disease. He asks, 'Does the use of animal models of disease take us any closer to understanding human disease? With rare exceptions, the answer to this question is likely to be negative.' He also criticizes assumptions made about in vitro tests for the same reasons above and argues that we need to get back to the human patient to truly understand human disease. Horrobin is correct in arguing for the need to study human patients. However, as will be discussed later, in vitro tests, if using *human* cells and tissues (not cells from another species) and if used in concert with other human-based testing methods, are more likely to accurately predict human outcomes than animal tests.

Horrobin is not alone in observing the incongruency between what we are studying in animals and what we should be studying. It is extremely troubling that because of our focus on animal models we know far more about a vast array of diseases in animals in the laboratory and how to treat them in animals than we do in humans (recall Christopher Reeve's comment).[68] In 2004 *New Scientist* reported on sentiments about the state of neuroscience research expressed by Susan Fitzpatrick, former Associate Executive Director of the Miami Project to Cure Paralysis and current Vice-President of the James S. McDonnell Foundation:

> 'The biomedical model is failing,' says Susan Fitzpatrick.... Basic biomedical research relies heavily on animal models, especially rats and mice, but she thinks it may be necessary to rethink this approach if treatments for brain diseases are going to reach the patients who need them. Even if we know all there is to know about the animal model we don't necessarily know about the disease, Fitzpatrick says. 'The model becomes what we study, not the human disease.'[69]

This sentiment can be applied to most human diseases. Rather than spending our time trying to unravel the mysteries behind human diseases directly, we instead create artificial animal models in the laboratory and these become our focus of attention. The *New Scientist* article continues:

> 'Take brain cancer. The traditional model for studying brain cancer is to take human cancer cells, sometimes tissue-cultured into cell lines, and transplant them under the skin of an immunosuppressed mouse. This approach ignores the fact that cancer is a disease of context: as soon as you change the environment you will change those cells. Any agent you test is probably unlikely to be effective when you have a tumour in context,' Fitzpatrick says. 'It's a fundamental flaw. We need a fundamentally new approach.'

Lost in translation: Species differences

Even when we think we have created an animal model that adequately mimics a human disease, interspecies differences come into play. In spinal cord injury, drug test results vary according to which species, and even which strain within a species, is used, largely because of numerous inter-species and inter-strain differences in neurophysiology, anatomy and behavior.[70] For example, the micropathology of spinal cord injury, injury repair mechanisms and recovery from injury vary greatly between different strains of rats and mice.[71] Surprisingly, even rats from the same strain but purchased from different suppliers produce different test results.[72] In one study, responses to 12 different behavioral measures on pain sensitivity, which is often used as a marker of spinal cord injury severity and recovery, varied among 11 strains of mice, with no clear-cut patterns that allowed prediction of how each strain would respond.[73] Each of these and numerous other differences influenced how the animals responded not only to spinal cord injury but also to any potential therapy being tested. A drug might help one strain of mice recover but not another.

There has been considerable enthusiasm for using mice as human disease models because of their ostensible genetic similarity with humans and because their entire genome has been mapped.[74] Mice have been extensively studied and, other than rats, are the most common animals used in experimentation. Scientists have modified their genes and created a host of new mouse strains designed to mimic a range of human diseases. Arguably, we know more about mouse physiology than we do about any other species, even humans. But do we know enough? In 2006, researchers reported in the journal *Science* the discovery that mice normally have more than one thymus gland.[75] Before this discovery, the predominant scientific view was that mice possessed only one. Since the thymus affects immune system function, experimenters had for decades been removing the one murine thymus gland of which they were aware, believing that they then created immune-deficient mice. Now we know the results of over half a century of research in immunodeficiency in thymectomized (thymus gland removed) mice were likely misleading. 'From the immunological point of view,' commented the study co-authors, 'a regular second thymus in mice raises important questions about previous studies using thymectomized mice.'

A 2006 report in the *Proceedings of the National Academy of Sciences* revealed that the internal structure of the human pancreas—including the insulin-producing Islet cells and surrounding cellular architecture—differs markedly from the experimental rodent models used for more than three decades.[76] Furthermore, these differences in architecture result in distinct differences in pancreatic function between mice and humans. The authors concluded that we cannot rely on mice and rat studies and that researchers must focus on human pancreatic cells and tissues. Perhaps most shockingly, this simple description of human pancreatic structure instantly invalidated decades of mice and rat experiments that relied on the assumption of similar pancreatic structure and function between humans and these animals.

These two examples of inaccurate assumptions about mice and rats are just the tip of the iceberg. Moreover, discovering a second thymus or comparing the cellular anatomy of mice and humans are relatively simple investigations to conduct and simple answers to confirm. How many other false assumptions are made because of questions we don't even know how to ask, yet alone answer? An article published in *Drug Discovery World*, entitled 'The importance of using human-based models in gene and drug discovery', noted that 'Mice and humans have more than 95% of their genes in common, yet mice are not men (or women).'[77] University of Michigan evolutionary biologists Ben-Yang Liao and Jianzhi Zhang found that although mice share most of their genome with humans, identical genes may behave very differently between the two species.[78] They compared human and mouse orthologs, which are genes in different species that evolved from a common ancestral gene. Normally, it is assumed that orthologs retain the same function in closely related species, such as mice and humans, during the course of evolution, and this assumption is a main basis for the use of animal models to study human biology.[79] Essential genes are those that, following loss of their function, reduce the fitness of an organism to zero. Liao and Zhang identified 120 human genes for which the mouse has an identical counterpart and discovered that 22 percent of the essential genes in humans are nonessential in mice. The authors concluded that 'it is possible that mouse models of a large number of human diseases will not yield sufficiently accurate information'. Commenting on this study, a scientist from the Dr Hadwen Trust, a medical research charity that funds the development of human-based testing methods, reflected, 'We have long been concerned that equivalent genes in humans and mice don't have the same functional effects. Millions of genetically modified mice are used as research "models" for human diseases every year but the relevance of this research to human patients is highly questionable.'[80]

A study at Massachusetts Institute of Technology demonstrated wide differences in the regulation of the same genes between the human and mouse liver.[81] Consistent phenotypes (observable physical or biochemical characteristics) are rarely obtained by modification of the same gene, even among different strains of mice.[82] Gene regulation can substantially differ among species and among individuals within a species and may be as important as the presence or absence of a specific gene. The disruption of a gene in one strain of mice may be lethal, whereas disruption of the exact same gene in another may have no detectable phenotypic effect.[83] Such findings question the wisdom of extrapolating data that are obtained in mice to other species. 'If one mouse gene is so difficult to understand in a mouse context,' asks Horrobin 'and if the genome of a different inbred strain of mouse has so much impact on the consequences of that single gene's expression, how unlikely is it that genetically modified mice are going to provide insights into complex gene interactions in the... human species?'[84]

'Humanized' mice

Genetically engineered mice are extensively used in amyotrophic lateral sclerosis (Lou Gehrig's disease) experiments but they are increasingly found to be inaccurate models of the disease and their use has failed to result in any effective treatment.[85] Cystic fibrosis knockout mice (genetically engineered mice in which one or more genes have been turned off through a targeted mutation) don't display the bronchopulmonary signs that are characteristic of human cystic fibrosis.[86] Despite their genetic similarity, there are fundamental differences between tumor cells in mice and humans. For example, in comparison with human tumor cells, those in mice tend to grow much more rapidly and are much more dependent on the formation of new blood vessels.[87]

Stanford University immunologist Mark Davis blames some of our limited understanding of the human immune system on our reliance on experimentation in mice.[88] As an example, he describes the results of tests using a type of protein to treat multiple sclerosis, an autoimmune disease: 'Injecting [myelin basic protein (MBP)] into mice causes a condition similar to multiple sclerosis, which can be prevented by doses of proteins that blunt the immune reaction to MBP. But clinical trials of these protective proteins were stopped because they made some people with multiple sclerosis worse.' A study published in *Science* in 2009 found that a crucial protein found in humans to regulate blood sugar is not found in mice, calling into question the relevance of the mouse model in the development of drugs to treat human diabetes, and suggesting that testing potential diabetes drugs in mice might give misleading results.[89] Even when the protein was expressed in genetically altered mice, it behaved differently than it does in humans. Genetic mouse models are poor substitutes for a number of other human conditions.[90] As we have seen with the multiple sclerosis trial example given above, reliance on mouse models has led to direct human harm. In 2003 Élan Pharmaceuticals had to stop trials of an AD vaccine that had cured the disease in 'Alzheimer's mice' after the substance caused brain inflammation in humans.[91]

The more we look into their effectiveness, the more we discover that genetically engineered animal models aren't living up to their promise. Perhaps the major and immutable reason genetically modified animals will not solve the problems of animal experimentation translation to humans is the fact that the 'humanized' genes are still in non-human animals. When we introduce a 'humanized gene' into a mouse, that gene will be affected by all of the physiologic mechanisms that are unique to the animal. As aptly stated in *Slate* magazine, 'tinkering with a few genes doesn't make [mice] perfect stand-ins for people'.[92] Short of turning mice into human beings, no matter how we modify their DNA there will always be significant disparities between their physiology and ours.

Do non-human primates make good models?

Drug testing regulations often require the testing of a new agent in both rodent and non-rodent species. Non-human primates (NHPs) are widely used as the non-rodent species. Yet NHPs, despite their even closer evolutionary history and genetic make-up to that of humans, also make far from ideal stand-ins for human-based tests. In March of 2006, six healthy human volunteers were injected with small doses of TGN 1412, an experimental therapy for rheumatoid arthritis and multiple sclerosis, created by TeGenero. As described by *Slate*,

> Within minutes, the human test subjects were writhing on the floor in agony. The compound was designed to dampen the immune response, but it had supercharged theirs, unleashing a cascade of chemicals that sent all six to the hospital. Several of the men suffered permanent organ damage, and one man's head swelled up so horribly that British tabloids refer to the case as the 'elephant man trial'.[93]

What went wrong? Were there too few animal experiments conducted prior to the clinical trial? No, TGN 1412 was tested in mice, rabbits, rats and monkeys with no ill effects.[94] Were the animals used not the appropriate animals to use? The answer to this also appears to be no. TeGenero intentionally selected cynomolgus monkeys for preclinical testing because they proved to best replicate a wide variety of mechanisms in humans specifically targeted by the drug.[95] Thus, not only were several different species used, but those deemed most relevant to humans were used. Did the problem then lie in the dose given to the test animals? Again the answer is no. Monkeys underwent repeat-dose toxicity studies and were actually administered 500× the dose given to the human volunteers for not less than four consecutive weeks.[96] Still, none of the monkeys manifested the ill effects that humans showed within minutes of receiving a minuscule amount of the test drug.

The problem with the TGN 1412 experiments is not that an inappropriate animal, dose or study design was used. The problem is that pharmaceutical research is now producing sophisticated, complex and nuanced molecules targeting very specific mechanisms in humans. Despite our close genetic relationship with NHPs, they are still not similar enough to make good models. In fact, humans are not always similar enough to other humans. We widely recognize that there are many differences in physiology and susceptibility to disease, and in effectiveness and side effects of treatments between individuals and groups within our own species. Hence, there is a growing interest in personalized medicine, in which treatments are tailored to individual patients. When clinical trials are conducted on a new blood pressure medication, for example, these, with rare exceptions, tend to include African-Americans, Hispanics, Asians and women because the results may vary between these groups. What works for a Caucasian male may not work for a Caucasian female or an Asian male. Scientists recognize the diversity in physiology within our own species, even among identical twins with the same genetic make-up. Twins display different susceptibility to diseases and genetic responses from one another and these responses become more disparate as the twins age.[97] If we can't reliably extrapolate from one identical twin to another, how can we expect to safely extrapolate results from completely different species to humans?

Our closest genetic cousins—chimpanzees—share about 95–96 percent of our genes but less of our DNA because of the tens of millions of differences in non-coding regions of our DNA. Many studies have demonstrated multiple disparities between chimpanzees and humans in DNA sequence, genetic insertion and deletion events, genetic expressions and post-translational modifications.[98] A recent study found a wide variety of both subtle and large-scale differences between chimpanzees and humans in cell death and DNA repair mechanisms.[99]

NHP models fail to reproduce key features of Parkinson's disease, both in function and in pathology.[100] Several therapies that appeared promising in both NHP and rat models of the disease showed disappointing results and even higher incidence of adverse effects in humans.[101] NHPs are not good severe acute respiratory syndrome (SARS) models either, even though an enormous undertaking has been made to reproduce the disease in them.[102] Long-time SARS researcher Robert Hogan recently argued against the further use of NHPs given that so many groups, such as the Centers for Disease Control and Prevention and the Army Medical Research Institute of Infectious Diseases, have reported contradictory results with SARS testing in NHPs.[103] Chimpanzees have been widely used to develop vaccines against hepatitis C under the presumption that they closely resemble humans in their response to the virus, despite the fact that the supporting evidence to this claim is slim.[104] After decades of this line of investigation we still have not developed any hepatitis C vaccine that works well in humans.

HIV/AIDS vaccine research using NHPs is probably one of the most notable failures of translation to humans. A lot of time and energy has been spent studying HIV in chimpanzees and other NHPs. In 2007, Alison Tonks, the associate editor of the *British Medical Journal* (*BMJ*), wrote about another failed HIV vaccine, gp120, and commented that important differences between monkey models and humans with HIV have misled researchers.[105] More than 85 HIV vaccines have failed in about 200 human trials following success in NHPs.[106] One of the most recent disappointments occurred in 2007 when a clinical trial testing a novel HIV vaccine developed by Merck (MRK-Ad5) was halted prematurely because it was actually found to increase the risk of HIV in certain groups of people.[107] MRK-Ad5, like all candidate HIV vaccines, was advanced into human trials after extensive preclinical experiments in NHPs.[108] The British newspaper the *Independent* summarized the incident as follows:

> One of the major conclusions to emerge from the failed clinical trial of the most promising prototype vaccine, manufactured by the drug company Merck, was that an important animal model used for more than a decade, testing HIV vaccines on monkeys before they are used on humans, does not in fact work.[109]

A recently published review found a paucity of evidence demonstrating successful translation of NHP research to human medicine in toxicology, stroke, AD, Parkinson's disease and infectious disease research.[110] It revealed that most data suggested experimentation on NHPs, including chimpanzees, to be irrelevant and unnecessary, to have little or no predictive value and to be hazardous to human health. For example, the campaign to prescribe hormone replacement therapy in thousands of women to prevent heart disease and stroke was based in large part on experiments on NHPs. Hormone replacement therapy is now known to *increase* the risk of these diseases in women. The bottom line is that despite assumptions to the contrary, the evidence tells us that NHPs simply don't reliably make effective models of human diseases.

Toxicity testing in animals

Of all fields in medicine involving animal experimentation, none is getting as much scrutiny as toxicity and carcinogenicity testing. One of the most extensively used methods to predict the carcinogenicity of a substance is the costly and time-consuming two-year bioassay in which mice and rats are exposed to maximum tolerated doses of test chemicals for two years to determine whether the chemicals are carcinogenic. Health agencies in the USA and abroad have hailed this bioassay as the 'gold standard' in carcinogen identification.[111] These accolades appear premature as the human relevancy of this testing method is becoming increasingly dubious.[112] A growing body of evidence suggests that some chemicals produce cancer in mice and rats through species-specific mechanisms that are irrelevant to human physiology.[113] For example, male rats get bladder cancer from saccharin through a rodent-specific mechanism (humans lack the protein that is necessary for the development of cancer in rats).[114] Based on this understanding of the species' differences, the NIH dropped saccharin from its list of human carcinogens in 2000. Phenobarbital is carcinogenic in rats because it raises levels of thyroid-stimulating hormone (TSH), which triggers thyroid cancer cell development.[115] But it does not substantially raise TSH in humans, if at all, so our cancer risk from the drug is negligible.

The false-positive and false-negative results of the animal bioassay can be considerable. Ennever and Lave analyzed the data on known human carcinogens with the animal data for cancer predictability.[116] They found a disturbingly large proportion of incorrect predictions, 'potentially allowing widespread human exposure to misidentified chemicals'. An analysis of the data on 780 chemical agents listed in the International Agency for Research in Cancer database found the positive predictivity of the animal bioassay for a definite or probable human carcinogen to be only around 20 percent.[117] In addition to placing human lives at risk, the low predictability of this assay is costing us money and wasting time. Each assay requires up to millions of dollars and years of planning.[118] In the meantime, as we continue to rely on this assay, there is a huge backlog of untested chemicals to which we are already exposing ourselves.[119]

Other toxicology and carcinogenicity tests that rely on animals are equally flawed. One study examined the toxicological profiles of 50 compounds in rodent and non-rodent (beagles and NHPs) species.[120] The study found poor correlation of target organ toxicity across species and concluded that 'simple extrapolation across species is unrealistic'. The study authors called for regulatory agencies to institute an evaluation of tests using animals as predictors of human adverse signs. In 1999 the Health and Environmental Science Institute examined the data on 150 compounds that had produced a variety of toxic effects in people.[121] It found that only 43 percent of the compounds produced similar effects in mice and rats and 63 percent did so in other animals. A reviewer of toxicology testing and regulations commented that

> compelled to act, regulators have chosen animal tests to forecast human cancer risks. To this end, animal data are filtered through a series of preconceived assumptions that are presumed to overcome a host of human/animal differences of biology, exposure, and statistics-differences that in reality are insurmountable.[122]

Recognizing the immense difficulty in predicting toxicity in one species based on the toxicity data from another is not new. As early as 1978, Fletcher found poor correlation between drug safety tests in animals and subsequent clinical experience with 45 major drugs, including anticancer agents, antibiotics, cardiac agents and neurological agents.[123] Fletcher's survey established that only 25 percent of the toxic effects observed in animals might be expected to occur in humans. Assessing three decades of data on the subject, toxicologist Ralph Heywood also found that the concordance between animals and humans is only 25 percent.[124] 'Toxicology,' he concluded, 'is a science without a scientific underpinning.'

'In retrospect,' Fletcher concluded in his 1978 report, 'it is a relatively simple matter to determine the correlation between animal and human studies, but prospectively it is difficult to know which particular toxic effects are likely to prove troublesome when it comes to giving the drug to man.'[125] And that's the catch: accurately predicting when the animal experimental results are relevant to humans is nearly impossible because of inter-species differences. We can always go (and have often gone) back after clinical trials have been conducted to assess whether the animal experimental results correlated with the clinical results, but retrospective confirmation is not the purported reason for using animals in experimentation. They are intended to predict human results and inform human health care. If we find that the animal experimental results equated with the clinical results, then the research community hails the efficacy of the animal experiments. But when the animal and human results do not match, the proclaimed failure is said to be a result of flaws in experimental design, publication bias or use of young animals for a disease that occurs predominately in elderly humans. Rarely is the use of the animals themselves—not how they are used—questioned.

While most researchers admit the difficulty in extrapolating and applying information obtained from other species to humans, commonly proposed solutions to this colossal obstacle are far from helpful. Neyt et al. suggest that 'clearly profound differences may exist at the gross, microscopic and genetic level between humans and other mammals, and these differences must be appreciated before extrapolating the results of a given study to human clinical practice'.[126] Caution in extrapolating data from animals to humans is another common advice given.[127] In fact, 'appreciation of differences' and 'caution' about extrapolating results from animals to humans are now almost universally expressed in published reports on animal experimental results intended to inform human health. Yet, in reality, how does one take into account differences in drug metabolism, genetics, expression of diseases, anatomy, behavior, influences of laboratory environments, and species and strain-specific physiologic mechanisms and then discern what is applicable to humans and what is not? There is just no established formula or algorithm to do this. Many scientists have recently acknowledged that modeling human disease in animals is extremely problematic but have still argued for their use, instead, to study basic physiologic mechanisms.[128] But again, if we cannot predetermine what mechanisms in what species and what strain of species and in what caging system and even during what time of day are applicable to humans, then the usefulness of the experiments needs to be questioned.

As reviewed earlier, basic research using animals is not effectively leading to new therapies to improve human health, which is the ultimate goal of medical research. A 2003 *American Journal of Medicine* review of 101 of the most heralded basic science discoveries from 1979 to 1983 revealed how unreliable even the 'cream of the crop' basic science findings can be when transferred to human medicine.[129] Following the course of these 101 breakthrough discoveries for up to 20 years, the authors found that only 27 resulted in published randomized clinical trials, only 5 were approved for human use and just 1 (a blood pressure drug) had a major clinical impact. The authors concluded, 'Even the most promising findings of basic research take a long time to translate into clinical experimentation, and adoption in clinical practice is rare.' Successful translation of basic research is, in fact, fairly uncommon.[130]

Of course, similarities in physiologic mechanisms exist across all species used in experiments and in humans. However, given the way medicine is practiced today, the differences between species appear to far outweigh the similarities and a growing body of evidence is attesting to this. The shortcomings of animal experiments for extrapolation to humans across a wide variety of fields are evident.[131] These include cancer, systemic sclerosis, osteomyelitis, asthma, Huntington's disease, Parkinson's disease, multiple sclerosis, fibromyalgia, alcohol addiction, sepsis (infection of the blood), shock and behavioral disease and psychiatric illness research.[132] Although only a few studies have systematically or critically reviewed whether animal experiments predict human outcomes, these are confirming the unreliability of animal experiments in a number of areas.[133]

Moving science forward

The argument that animal experiments are largely unreliable predictors of human disease mechanisms and health outcomes does not dismiss the fact that some animal experiments have proved successful. Statistically, it is inevitable that some animal experimental results will match human results. As Michael Bracken from the Yale School of Public Health stated, 'given the large number of animal studies conducted, it would be expected that some animal experiments do predict some human reactions'.[134] Based on these successful examples, one may argue that, despite the many limitations, animal experiments have provided useful information. While this assertion would certainly not be inaccurate, the question remains: is animal experimentation the *best* way to get the information we need *today?* The earliest telescopes gave us a glimpse of the universe around us, but they lacked the accuracy for us to target and discern the critical details that would allow us to arrive at a more comprehensive understanding of how the universe functions. Similarly, although animal experimentation may be one means by which we gain some understanding of physiologic and disease mechanisms, the details of these mechanisms that are human-specific and relevant to human health too frequently remain a mystery. Thus we are left with, at best, incomplete and, at worst, inaccurate information.

Even if animal experiments are causally related to the production of data relevant to human health, it does not follow that animal experiments are the only, or even the most efficient, way to obtain relevant data.[135] We are just starting to recognize how minor variations between species can substantially perturb study results. These are just the variations of which we are currently aware. They do not include the many differences between species and strains within a species that we have not yet discovered. These known and likely far more unknown differences render it extremely difficult to unravel and determine what results, if any, from an animal experiment can or cannot be applied to humans. The pivotal argument against using animals as models of disease or to study basic mechanisms is that it is impossible to know in advance which models and which mechanisms will show the same results as in humans. Evidence that some animal experiments accurately predict human results or provide useful information does not detract from the many costly and devastating failures or refute the underlying premise that extrapolation from animals to humans is highly tenuous.

It has been argued that some information obtained from animal experiments is better than no information.[136] This neglects several crucial points that illustrate how a little knowledge can be a bad thing, especially if it is dubious. As we have seen with some of the examples presented, many people have been directly, and often significantly, harmed because researchers were misled by the safety profile of a new drug based on animal experiments. A large number of people volunteering in clinical trials have put their lives at risk based on animal experimental results, which often turned out to be inapplicable to humans. A review in the *BMJ* expressed it thus: 'Biased or imprecise results from animal experiments may result in clinical trials of biologically inert or even harmful substances, thus exposing patients to unnecessary risk and wasting scarce research resources.'[137] We may already be exposing ourselves to numerous carcinogenic chemicals because animal tests were falsely negative. Thus, far from protecting us, animal experimentation often puts us at greater risk.

Furthermore, the indirect human harms caused by the opportunity costs may be substantial. An invalid disease model can lead the industry in the wrong direction, wasting time and significant investment.[138] Repeatedly, researchers have been lured down the wrong line of investigation because of information gleaned from animal experiments that later proved to be inaccurate, irrelevant or discordant with human biology. It's taken more than 25 years of failed HIV vaccine clinical trials for researchers to seriously question the usefulness of NHP HIV models, and more than 30 years before we realized that the rodent model of diabetes is wrong. A substantial amount of human suffering could have been prevented if instead we had focused on studying HIV and diabetes solely through human-based tests.

Treatments that fail to work or are harmful in animals may be effective and safe in people. Robert Wall and Moshe Shani from the USDA wrote that

> it is interesting to speculate that animal models may be just as likely to exhibit false positive results (compound or devise would be OK in humans but show adverse effects in animal studies) as they do false negatives results (OK in animal studies but have adverse outcomes in human trials).[139]

Animal experimental results may have caused us to abandon countless therapies, which could have worked in humans and alleviated untold suffering. Of every 100,000 chemicals tested in the lab, only about 50 pass on to phase 1 clinical trials. Most don't show enough benefit, aren't easily absorbed in the body or are harmful to animals.[140] But many of these agents may have worked spectacularly in humans.

Aspirin is considered one of the best drugs we have today, despite the fact that its discovery took place over 100 years ago. A recent report examined the safety profile of aspirin in experimental animals.[141] The results showed that in different animal species, aspirin is a cancer promoter, 'harmful if swallowed', a 'respiratory irritant' and causes other serious adverse effects. The report concluded that we are extremely fortunate that we did not rely on animal experiments in 1899 to decide whether to approve aspirin for use in humans by saying 'it is not very likely that any substance with such a profile would make it to clinical trials or to the market today'. This holds true for many well-known drugs, including acetaminophen. Experiments on animals delayed the acceptance of cyclosporine, and Fk-506 (tacrolimus) was almost shelved because of high toxicity in animal experiments.[142] Both drugs are widely and successfully used to treat autoimmune disorders and prevent organ transplant rejection in people. Experiments on mice provided no evidence whatsoever of the efficacy of beta-agonist bronchodilators in the treatment of asthma and suggested that thiazolidinedione anti-diabetes drugs would actually make diabetes worse, in contrast to human studies.[143] A report in *Slate* magazine rightly noted that 'an equal source of human suffering may be the dozens of promising drugs that get shelved when they cause problems in animals that may not be relevant for humans'.[144]

The costs to animals

In addition to causing direct and indirect human suffering, reliance on animal experimentation causes a vastly underappreciated amount of pain and suffering in animals. Annually, more than 115 million animals—including mice, rats, frogs, dogs, cats, rabbits, hamsters, guinea pigs, monkeys and birds—are used in experimentation or bred to supply the research industry worldwide, many of whom endure intense suffering. Approximately 42 percent of NIH-funded research involves experimentation on animals.[145] That translates to more than $12 billion spent on animal experimentation in 2009 alone in the USA, not including the substantial amount coming from the pharmaceutical sector and other governmental and private entities.[146] In the USA in 2009, more than 76,000 animals were subjected to pain without being provided with pain relief.[147] This number does not include the majority of animals used in experimentation (rats and mice), birds, reptiles, amphibians and most animals used in agricultural experiments, all of whom are excluded because they are not considered animals under the Animal Welfare Act (AWA).[148] There are no federal requirements to report the number of these animals used in experimentation or the types of procedures conducted on them. Thus potentially hundreds of thousands of animals may be subjected to painful experiments annually without being provided with any pain relief at all.

In Canada, more than 3 million animals were used in research, teaching, testing and the production of biological products in 2009, an increase from prior years.[149] More than 145,000 were subjected to 'severe pain near, at, or above the pain tolerance threshold of unanaesthetized conscious animals'. The number of animals subjected to this severe pain increased from 55,000 in 1998. At least 11 million animals are used each year in experiments in the European Union.[150]

With rare exceptions, scientific interest always trumps the welfare of the animals. The US *Guide for the Care and Use of Laboratory Animals* stipulates there should be 'proper use of animals, including the avoidance or minimization of discomfort, distress, and pain', but, and this is the important point, '*when consistent with sound scientific practices*' (emphasis added).[151] Thus, the scientific endeavor overrides animal welfare concerns, even for those animals covered by the AWA. All experimentation, no matter the level of pain and suffering, is potentially justifiable by these guidelines. As one bioethicist notes, 'Of particular importance, the appeal to animal welfare in the regulatory guidelines avoids any commitment to limits on what can be done to animals for the sake of human interests.'[152] Other regulatory guidelines in the USA and abroad are severely deficient in protecting animals from harm.[153] As an example, the AWA does not set forth any standards by which animals are to be kept but leaves that to the USDA.[154] Marian Sullivan, Deputy Chief Court Attorney at the New York State Supreme Court, explains that the AWA requires the USDA to set forth humane care standards. Essentially, however, 'the standards set forth by the USDA...require little more than that animals be fed, watered, vetted, and kept in reasonably clean and safe enclosures that allow them to make species-appropriate postural adjustments'.[155] In other words, the AWA is basically a husbandry law that stipulates that animals be fed and be allowed to move about somewhat in their cages.

Ultimately, anything can, and arguably has, be done to animals in the laboratory setting. Every year we poison, bludgeon, shoot, crush, gas, infect, drown, blind, dismember, burn and electrically shock animals in the name of research—often without any pain relief. A survey was recently conducted by one of the top authorities of analgesic use in animals, Paul Flecknell. He found that of the published papers that reported the use of mice or rats in extremely painful, invasive procedures such as burn experiments, spinal cord injury experiments and skull surgeries, post-procedural pain relief was provided to the animals only 20 percent of the time.[156] Moreover, an estimated 50–60 percent of mice and rats receive no pain relief whatsoever both during and after the painful procedures. Signs of psychological distress, including stereotypic or repetitive movements, self-injurious behaviors, near catatonia, vocalizations, inappropriate aggression, fear or withdrawal are all commonly seen in animals in the laboratory.[157] About half of all mice used in experiments are estimated to be afflicted with behavioral stereoptypies.[158]

Earlier in this chapter, studies were presented which demonstrated that animals respond to routine laboratory procedures, such as handling and blood collection, with rapid, pronounced and statistically significant elevations in stress-related markers. Common responses by NHPs to routine procedures include fear grinning, vocalizations, diarrhea and physical resistance (such as struggling or refusing to enter a cage).[159] The simple act of catching an animal and removing him from his cage can cause significant elevation of his plasma cortisone levels.[160] Several studies in monkeys, mice and rats suggest that witnessing other individuals being subjected to unpleasant laboratory procedures is stressful.[161] Animals watching their cagemates being captured for a procedure are affected by 'contagious anxiety'.[162] Rats show significant elevations in heart rate and blood pressure when present during decapitation of other rats.[163] Cortisol levels shoot up in monkeys able to see other monkeys being restrained and sedated for blood collection.[164]

These findings suggest that the responses in animals to the laboratory procedures are more than mere arousal responses and are indicative of stress and distress.[165] One study found that when an individual in a laboratory coat with a catching net entered the room where monkeys were housed, the monkeys displayed substantial expressions of negative emotion and changes in body temperature indicative of distress.[166] What this and other such studies demonstrate is that animals do not readily habituate to the laboratory environment or procedures; they just don't get used to it. Fear and anxiety are daily phenomena of their lives. Even if we try to make life a little easier for these animals, by housing 'enrichment' and by more routine use of pain medications, for example, ultimately we just cannot get around the fact that the laboratory settings, daily procedures and experiments themselves cause tremendous suffering.

Despite the meager regulations covering only a minority of animals used in experiments, enforcement of even these is pitiful.[167] In the USA, the USDA's Animal and Plant Health Inspection Service (APHIS) is charged with overseeing the AWA. In 2005 the Office of the Inspector General published a scathing report of the USDA's failure to enforce the AWA.[168] It cited APHIS for not pursuing enforcement actions against violators, including repeat offenders, failing to effectively monitor research facilities, and charging minimal fees to violators. The report further concluded that the fines against violators were so minimal that 'violators now consider the monetary stipulation as a normal cost of conducting business rather than as a deterrent for violating the AWA'.

The new gold standard: Human-based tests

The last critical point against the argument that gleaning some information from animal experiments is better than none at all is that this argument assumes there is no alternative means of gaining medical knowledge. In addition to this being a false assumption, there is an array of proven alternative methods of testing that are in wide use today that reveal that we can gain *better* knowledge by not using animals. Sophisticated in vitro tests, human skin models for corrosion tests, genetic techniques, population studies, modeling methods, virtual whole-human modeling, virtual clinical trials, three-dimensional cell and tissue cultures, organs on a chip and imaging studies (using magnetic resonance imaging (MRI), functional MRI and positron emission tomography scans) are just a few examples of human-based testing methods currently available. Microdosing provides information on how an experimental drug is metabolized and its bioavailability throughout the human body. By administering an extremely small dose (i.e. well below the threshold necessary for any potential pharmacologic, and thus harmful, effect to take place), microdosing can be used safely in human volunteers.

Currently, many of these testing methods are being used in conjunction with animal experiments prior to the conduction of clinical trials. The problem with using both human-based and animal experiments, however, is that the latter may contradict findings from the former. When this occurs, as is often the case, the animal experimental results may be incorrectly favored (leading researchers down the wrong path of investigation) because they represent 'whole animal system' results. However, the animal tests provide the wrong whole systems. For genetic and physiologic reasons that are immutable, animal experiments are less trustworthy than even incomplete systems of the human body.

Some have argued that in vitro or other similar testing methods are simplistic and cannot accurately mimic the complexities of the human body, hence the need for animal experiments. In vitro tests certainly are prone to some of the same problems as animal experiments in that they can be relatively simplistic models of disease or physiologic mechanisms and are not always accurate. But are the animal experiments necessarily *more* accurate or predictive? A multicenter team of researchers evaluated 68 different methods to predict the toxicity of 50 different chemicals.[169] The animal tests were only 59 percent accurate, but a combined human cell in vitro test was 83 percent accurate in predicting actual human toxicity. Human skin cultured cells outperformed live rabbit tests in detecting chemical skin irritants. Tests in rabbits misclassified 10 out of 25 chemical irritants, while the cultured cells classified all irritants correctly.[170] Researchers compared in vitro human tumor cell lines with mouse cancer models for their reliability in predicting clinical phase 2 trial results of 31 potential cancer drugs. The study found that the in vitro tests were reliable in predicting the clinical utility of these drugs for all four cancer types tested, whereas the mouse allograft cancer model (in which cancerous tissue from one mouse is transplanted into another) was not predictive.[171] The human xenograft mouse model (in which cancerous tissue from a human is transplanted into a mouse) was predictive for only two of the four cancer types studied. The study authors concluded that cancer drug development emphasis should be placed on in vitro cell lines.

An in vitro test developed by UK researchers could have predicted TGN 1412's serious adverse effects before it was ever tested on
humans.[172] In all of these examples, 'test tube' experiments were far
more accurate than whole animal model systems. Asterand has confirmed that studies on human bronchial smooth muscle and pancreatic
Islets tissues are far better at predicting human responses to asthmatic
and diabetic drugs than animal experiments.[173] One of the best features
of in vitro methods is that we have better control and understanding of
the testing parameters. With animal experiments, especially because
of the many inter-species differences in physiology of which we are
not aware, our control and understanding of the testing influences are
greatly limited in comparison. Regardless of the preference to study
whole biological systems, non-human animals are not the correct systems. An understanding of human physiology is critical. And it cannot
be overstated that in order to be the most accurate and predictive
as possible, in vitro tests must use *human* cells and tissues, not cells
from other species, otherwise interspecies differences still come into
play. While there is no perfect predictive approach to human medicine,
a combination of human-based testing methods, including in vitro
tests, will likely get us closer to the true answers than animal experiments, which are inherently flawed. Human-based in vitro tests may
not always be accurate predictors of human responses, but they have
great potential to become more accurate, particularly as new methods
are developed that are closer to depicting whole human systems. At a
fundamental level, non-human models, on the other hand, cannot be
accurate, and cannot be made to be accurate, because of distinctions in
genetic make-up and expression, and evolutionary issues such as causal
disanalogy.

Biotechnology company Selventa (formerly called Genstruct) has
compared animal models of human disease with the actual human
diseases.[174] For instance, it has studied mouse and rat models of type
2 diabetes. In many cases, it found that other than having aberrations
in insulin signaling and glucose levels, there was no similarity between
the animal model and the human disease condition. 'If you're developing a drug in that animal model, it's clearly not going to work in humans
because they have a different disease,' says Keith O. Elliston, Selventa's
president and chief executive officer. Selventa focuses on human-based
tests. These are far from simplistic and deal with the complexity of
human biological systems. It has developed in vitro models that include
all the genes, proteins and metabolites present in human cells. The company then applies artificial intelligence tools to work through all the

predicted and observed relationships among these components and put them in context within the complex system.

Many other forward-thinking companies are exploring modern alternatives. Pharmagene Laboratories, based in Royston, UK, is the first company to use only human tissues and sophisticated computer technology in the process of drug development and testing; it does not conduct any animal experiments.[175] With tools from molecular biology, biochemistry and analytical pharmacology, Pharmagene conducts extensive studies of human genes and how drugs affect those genes or the proteins they make. One of the co-founders asked, 'If you have information on human genes, what's the point of going back to animals?'[176]

Neurologists and other neuroscientists collaborating on the Miami Project to Cure Paralysis are using cutting-edge science to model human spinal cord injury.[177] Studying spinal cord-injured patients, researchers are gleaning a more complete understanding of human spinal cord injury. For example, they are comparing postmortem spinal cord tissue with MRIs of living patients to determine what changes in cells and tissues are detrimental. The project correlates neurological function, neurophysiology and findings from imaging studies and tissue pathology to design targeted therapies to improve the quality of life of injured patients and prevent further damage after acute injury. After only a few years, this project has made several notable discoveries about human spinal cord injury that were not made through animal experiments. It is the first project to provide evidence that humans possess specialized nerve circuitry that influences walking and could possibly be enhanced by rehabilitation training. It is also the first to show conclusive evidence of a critical neurological feature, chronic demyelination (disruption of the nerve coating necessary for proper nerve signaling) after spinal cord injury in humans, and to conceive and develop a novel intra-operative monitoring technique that makes spinal surgery safer.

In response to the limitations of animal immunology experiments for human health research, scientists at Stanford University are working on a 'Human Immunology Project'.[178] The investigators are using high-throughput screens to catalog a host of cellular parameters.[179] They are using a systems biology approach to understand the many facets of the human immune system and how the whole system fits together. Researchers have now created a virtual model of all the biochemical reactions that occur in human cells.[180] A major report released in 2007 by the National Academies' National Research Council (NRC) called for a transformation in toxicology testing—one that largely shifts away from animal experiments.[181] The NRC recommended the development and use of in vitro methods using human cells, in combination with computer modeling and other testing techniques, to evaluate changes in biologic processes and markers that would indicate toxicological effects in the human body. It concluded that not only would these new testing methods be more evidenced-based but they would also save significant resources and time in comparison with animal toxicity experiments.

These few examples of human-based testing methods are just a tiny sample of the sophisticated non-animal approaches currently available. Human-based methods must be validated and the ones that have undergone validation thus far are largely proving to be better than animal experiments in predicting human responses. While gaining momentum, human-based tests are still in their infancy and there are many areas in medicine where these methods need further development. This fact has been used to argue for the continued use of animal experimentation. But not having a viable alternative is not sufficient justification for continuing a misguided research paradigm. Instead, this line of thinking prevents us from any true commitment to finding new or improving existing alternative testing methods. It will cause us to continue to waste years and precious research dollars on sub-par methods, place humans at risk, cause suffering in animals, with perhaps the greatest tragedy of all being that we would likely abandon therapies that would have been effective.

Financial investments in the study of alternative testing methods pale in comparison with investments in animal experimentation.[182] For example, the US Government's agencies have spent less than $10 million over a ten-year period on validating alternatives for regulatory use, and validating alternative methods is rarely a priority for government funding.[183] The development of human-based alternatives to animal research is an underdeveloped field largely because so few resources are devoted to its development as a result of our commitment to animal-based methods.[184] Another major hurdle to the development and use of non-animal testing methods is that government regulations tend to require far more validation than was ever required, if at all, for the animal experimental methods they are intended to replace. Ironically, these new methods are often required to be validated against existing animal experimental methods, most of which have never been validated themselves.[185] This creates a double standard that allows the acceptance of most animal experimental methods as the 'gold standards' (based on tradition, rather than proven efficacy), providing a disincentive to the development of alternative methods.

An even larger problem with policies requiring the validation of human-based tests against animal experiments is that the latter are unlikely to predict human responses consistently, and may not even be consistent in general. Thus a human-based model might actually be consistent and predict human responses but would fail validation, while it is the animal test that is in fact inferior. Additionally, a final hurdle is that regulatory agencies do not usually mandate the use of alternative testing methods, where they exist and have been proven valid, in place of the traditional animal experiments. Thus, there is little incentive for pharmaceutical companies and others to switch gears and use alternative methods in research and drug development if they are already wedded to an animal model. Arguably, there has been a net loss of ground because alternative human-based methods, which would have likely gotten us further scientifically, have been neglected in favor of animal experimentation. It is time for this to change. It is incumbent upon investigators and research-supporting institutions to prioritize the replacement of animals in experiments. Failing to do so means delaying the development of more effective and accurate research techniques that could save thousands or millions of human and animal lives.

Dubious experiments we can eliminate

In the short term, we can agree that many experiments currently being conducted could be eliminated today. Consensus can be reached that a substantial proportion of animal experiments are highly irrelevant to human health. A quick exploration of some recently funded animal experiments attests to this. A survey of experiments conducted at US universities that were funded by the NIH was conducted in 2008 through the use of two databases: the Computer Retrieval of Information on Scientific Projects (CRISP), maintained by the NIH, and the CRISP*er* database, maintained by the non-profit Sunshine Project.[186] A literature review provided additional information. Examples of experiments funded by public tax dollars include:

> An experiment conducted between 2006 and 2007 by Emory University School of Medicine cost more than $97,000. In this experiment, muscle-recording electrodes were placed in anesthetized cats' hindlimb muscles. The cats were positioned over a treadmill, with their heads fixed in stereotaxic frames. Their brainstems were then cut and all brain matter above the incision removed. Anesthesia

was then eliminated and, as the cats initiated spontaneous stepping movements, the treadmill was turned on and muscle activity was recorded while the cats' heads were positioned in three different ways. The results were compared with results from intact cats (cats with brains intact). The main results suggest that modifying head pitch in a walking decerebrate (cerebral brain removed or disconnected) cat causes significant muscle activity changes that are similar to what occurs in an intact cat.[187]

At the Keck School of Medicine in southern California, an area of the frontal brain necessary for the sense of smell was removed through aspiration in male hamsters. The hamsters were then tested for their sexual attraction to male versus female hamsters. The goal of this experiment was to assess if, and how, testosterone, sexual experience and chemosensory cues play a role in sexual motivation in male hamsters. Between 1997 and 2006, this and similar experiments cost more than $1.8 million.[188]

At the University of Washington, sparrows were caught from the wild and deafened by puncture of their tympanic membranes. Their song production was then measured. The primary goal of this experiment was to assess whether deafening sparrows affected their singing and the seasonal growth of their song nuclei. This and other similar experiments on sparrows cost the public more than $3.4 million between 1997 and 2007.[189]

A series of mating behavior experiments on ferrets at Boston University between 1998 and 2007 cost more than $4 million. One of the major findings suggested that damage by electrical lesions to both sides of a part of the hypothalamus in the brain causes male ferrets to display a preference for sexual and body odors from other males over females.[190]

Experiments conducted at the University of California in which rats received repeated electric shocks revealed that as a defensive mechanism against a perceived threat, rats will hide and freeze in a familiar enclosure. This and similar experiments cost the public more than $8.6 million between 1997 and 2007.[191]

Between 1997 and 2007, the University of Michigan spent more than $21 million on experiments to assess whether alcohol reinforces the use of other drugs in monkeys.[192]

The examples presented here are far from isolated cases. Public funds are used to support numerous dubious experiments at medical centers and universities throughout the USA and abroad, regardless of their lack of relevance to human health. There are much better ways to use our tax dollars to improve human health than the examples above. Rather than continuing to pour millions of dollars each year into experiments on drug and alcohol use in animals, we could instead fund treatment centers for drug abusers. Rather than studying the song nuclei in sparrows, a nucleus that humans don't even have, we could instead fund experiments such as functional MRI studies of the changes in various areas of the brain in humans with deafness. Why not divert more funding to studies on human spinal cord injury, such as the Miami Project, rather than remove the brains of cats to monitor their spontaneous stepping activity, especially when humans, unlike cats, have little to no spontaneous stepping activity without input from the higher brain? Support for these experiments will inevitably revolve around suggestions that they will help elucidate underlying physiologic mechanisms that will one day have human health applicability. However, such a connection is extremely doubtful as we have seen how underlying mechanisms can differ so vastly between species. On the other hand, there is no doubt that there are many other ways to use these funds, which will benefit humans and will do so without causing animals harm.

Someone might claim that we don't know what benefit animal experiments, particularly basic research, may provide down the road. But as bioethicist Bernard Rollin pointed out, 'if that were a legitimate point, we could not discriminate between funding research likely to produce benefits and that unlikely to do so; however, we do. If we appeal to unknown but possible benefits, we are literally forced to fund everything, which we do not.'[193] Many researchers and funding institutions are aware of the fact that basic research on animals has come under intense criticism because society has hinted that that there are limits to what it would fund in terms of knowledge for the sake of knowledge. Consequently, much basic research on animals is now conducted under the guise of applied research.[194] However, as demonstrated in this chapter, the usefulness of basic research on animals to produce medical treatments is highly questionable. And we have seen how so few of even the most highly regarded studies in basic research ever translate to human benefit. Given the highly questionable usefulness and the immense suffering animals in laboratories experience, the appeal to serendipity in research is insufficient to justify an animal experiment.

Steps we can take

Regardless of the rationalizations given to support the use of animal experiments, the final test of their success is whether or not they improve our health and lead to new, effective, treatments or preventions. In this, they are largely disappointing. Animal experiments are proving to be extremely unreliable in predicting human outcomes. Is this a risk we want to continue to take? In drug production, the failure rate is at least 92 percent. More of us need to ask why we are failing so often. A 92 percent failure rate of *anything* should be cause for alarm. A 92 percent failure rate in drug development should likewise be unacceptable. Because the practice of animal experimentation is so entrenched in our current research paradigm, scientists who question the foundational relevance of animal experiments are often marginalized within the scientific community. Alternative opinions and studies critically examining the relevance of animal experiments are rarely, with some of the exceptions provided in this chapter, given an opportunity to be published in the biomedical literature. The failure by the scientific community as a whole to tolerate different opinions and publish critical examination of animal experiments is contrary to the very spirit of science and is a major obstacle to the advancement of human health.

Moving away from animal experimentation will no doubt take time. There are steps we can take today, though, to move us in a positive direction: a direction that is immensely beneficial to humans and animals, and that embraces more sophisticated and accurate testing methods. A thorough examination of how we spend our research dollars and the relevance of animal experiments to human health is vital. The public deserves accountability for how we spend their money. These steps require greater transparency in animal experimentation so that the evaluation of the experiments' human relevance and accurate assessment of the costs to both humans and animals can be made. They should include:

1. prioritization of the conduction and publication of critical and systematic studies evaluating the human health relevancy of animal experiments;
2. identification and immediate replacement of animal experiments agreed to be highly irrelevant to human health;
3. provision of transparency and registration of all animal experiments conducted by public and private institutions similar to clinical trials registries (such a registry should include the numbers and types

of animals used, details about the health and welfare of the animals, funding amounts, housing procedures and details on the experimental procedures conducted);

4. demand for a serious and primary dedication to development of non-animal testing methods; and
5. mandate the use of validated non-animal alternatives that currently exist in place of animal experiments.

We owe it to the public to use the best possible research methods. These are human-based tests. Their use also has the added benefit of avoiding the use of animals in harmful experiments. All we need is the willingness to question our own assumptions and the dedication to follow where this leads us. By doing this we will create a new gold standard for medical research—one based on sound science.

7

The New Public Health

The fate of animals is of greater importance to me than the fear of appearing ridiculous; it is indissolubly connected with the fate of men.

—Emile Zola

When Dr Albert Schweitzer accepted the 1952 Nobel Prize for Peace, he delivered his Nobel speech entitled 'The Problem of Peace', considered one of the greatest speeches ever given. At the time, the world had just passed through the Second World War and witnessed unfathomable cruelty, bigotry and injustice. In his speech, Schweitzer described the state of humankind that he believed allowed for so much suffering and declared that because of our command of science and technology 'Man has become superman.'[1] He stated, however, that 'superman' suffers from a fatal flaw—the failure to rise to 'superman' reason. This failure prevents us from applying our knowledge to useful ends, rather than to those that cause harm. As a result, that knowledge and power become a danger to us rather than an asset. In order to overcome our fatal flaw, Schweitzer challenged humankind to embrace its ethical spirit, which 'does not assume its true proportions until it embraces not only man but every living being'. He was primarily speaking of war, technological advancements in warfare and the pathway to peace, but his sentiments apply equally well to our health and our treatment of animals.

We have gained tremendous scientific knowledge and technological advancements that are allowing us to better control our own lives as well as the lives of billions of other animals. How we use our knowledge and scientific advancements to interact with animals affects our own welfare. Mostly we have used this knowledge, whether explicitly or implicitly, to industrialize and expand the suffering of animals, putting our own health at risk. We have used our knowledge to mechanize animal farms, to devise new methods of experimenting on animals and to collect menageries of animals for trading, while concurrently destroying their natural habitats. As explored throughout this book, however, the interests of human health are best served if we use our knowledge to help improve the treatment of animals and reduce their use in harmful practices. Public health is best served if we use human-specific testing methods in lieu of animal experiments, if we recognize animal abuse as a public health issue and if we reduce our use of animals as food and entertainment. In short, human health would be tremendously improved through better treatment of animals.

Partnering with police and humane organizations to investigate animal abuse may help us uncover children or domestic partners being abused and help prevent their further victimization. We can help victims of domestic assault—who stay with abusive partners for fear of harm coming to their companion animals—by ensuring that all victims of abuse, humans and non-humans, are provided with safe havens. And best of all, by identifying and catching animal abusers early on, we can help thwart future acts of violence against society.

Working with wildlife and animal protection organizations, we can help promote sustainable means of income and food for indigenous populations as alternatives to the bushmeat trade. This tactic has the benefit of boosting both the income and the health of such populations. We may be able to help protect them from catching novel and potentially deadly zoonotic pathogens similar to HIV and Ebola. We can help prevent pathogens from traversing the globe and infecting all of us by closing down live animal markets, encouraging the public not to purchase wild animals as pets, eschewing animal skins and furs as clothing, and seeking alternate means of entertainment that do not harm animals. Restricting or banning the keeping of wild animals as pets or for entertainment will also protect the public from deadly animal attacks.

Encouraging the public to combat factory farms with their forks and consume more plant-based diets will confer numerous health benefits. This will help lengthen our lives, keep us trim and prevent millions from suffering from stroke, heart disease, cancer and other chronic diseases. Our environment will be cleaner. Our food and water will be safer. The risk of deadly pandemics will be greatly reduced.

Promoting the development and use of ethical human-based testing methods will help us bypass the inherent and immense interspecies obstacles that arise from using animals in experimentation. More accurate human-based tests will give us greater confidence that the chemicals we use will not make us sick and that our medical treatments work and are safe. The *Washington Post* recently reported that the US Environmental Protection Agency has sufficient health and safety data for only 200 of the 84,000 chemicals in commerce.[2] Because the majority of chemicals on the market have not been sufficiently tested for their safety, major hurdles exist for the regulation of these products. In response to the lack of regulatory oversight, Wal-Mart and other retailers have taken it upon themselves to ban the sale of several chemicals in their stores and it appears that this 'retail regulation' will expand. Here is an example of others taking the lead where public health is failing. If we were to fast-track the development and use of high-throughput, human-based in vitro testing methods, we could more efficiently and more quickly test the safety of the thousands of chemicals currently in use. By doing so, we would best ensure that the use of unsafe chemicals is restricted and that we don't unnecessarily remove useful, non-harmful chemicals. Using existing human-based testing methods in place of animal experiments, such as three-dimensional human tissue cultures in combination with other combination with virtual clinical trials, human modeling, other in silico tests and imaging techniques, will help us develop medical treatments more effectively and with fewer disappointments. Lastly, rather than relying on problematic animal experiments, which may cause us to abandon promising drugs due to misleading results, we should substantially increase our investment in the development of further human-based tests that will better lead us down the road to improved medical treatments.

In addition to boosting our own health and welfare, the changes mentioned above are also beneficial to animals. Using more human-based testing methods will reduce the numbers of animals used in harmful experiments. Consuming fewer animals will reduce the numbers confined in factory farms. Restricting the global trade in wildlife will help prevent further destruction of animal habitats and protect many of them from abuse.

One could argue that the above changes can be made without considering the welfare of animals as these changes make sense for our own benefit. But, as is evident by the continuation and, in most cases, expansion of these harmful practices, these will not likely be significantly altered until there is a fundamental transformation in how we view animals. It is widely recognized, for example, that a major catalyst behind the drive to develop improved human-based tests is animal welfare concerns. Despite this drive, however, the replacement of the use of animals is still given low priority in medical research because animal welfare is not uniformly appreciated.

Humans will benefit if the current paradigm of our relationship with animals is replaced by one in which we no longer view animals as beneath our moral consideration. We are best served if we view animals not only for what they can do for us but also as beings with their own self-interests and inherent worth. This is because it is extremely doubtful that we will accomplish any real change in how animals are treated until we view animal welfare as an issue in its own right. Otherwise, the risk is that we will invariably view their welfare as secondary to other issues and will continuously devise new rationalizations to continue our current practices, even in the face of the enormous health benefits that can result from altering our practices. As long as we continue to view animals as tools and commodities, we will not successfully limit and do away with factory farms and the wildlife trade. We will not put any significant effort into expediting the replacement of animals in experiments and developing other testing methods. We will not strive to uncover abuses of animals. And if we cannot achieve these things, we will not see our risks of infectious diseases dramatically lowered, help thwart the cycle of violence, limit the destruction of our environment, and significantly improve our efforts to produce safer and more effective medical treatments. Thus, treating animals as ends in themselves and not just as means to an end is better for them and better for us.

Change is not easy and changing the current public health paradigm for one that recognizes the welfare of animals as an issue in its own right will take time and commitment. But at its best, public health has advocated social change and fought on behalf of the most vulnerable people in society, such as the poor, the socially marginalized, and children and adults with mental incapacitation and diminished decision-making capabilities. Animals are no less vulnerable, or perhaps they are even more vulnerable as we do not always recognize their vulnerability. Animals, like the most exploitable of humans, are utterly powerless against what we choose to do to them. We need to view animals as vulnerable to abuses in the same light we view other vulnerable people. The growing body of evidence affirms that animals feel pain and suffer—and at our own hands. As with us, animals wish to avoid pain, hunger, terror, loneliness and limitations on their freedom.

Kwame Anthony Appiah, philosophy professor at Princeton University, recently published an article in the *Washington Post* entitled 'What will future generations condemn us for?'[3] He asked his readers to consider how future generations will judge our current practices, given that we can now judge prior generations for the injustices they condoned. Appiah proposed four contenders for future moral condemnation: our disrespectful treatment of the elderly, our poor treatment of prisoners, our 'wasteful' attitude toward our planet's natural resources and factory farming. Regarding the deplorable conditions on factory farms, 'picture it', he states, 'and then imagine your grandchildren seeing that picture'. We can look beyond factory farms and picture all our current uses of animals and ask whether future generations will look back and shake their heads at our failure to recognize the obvious insensitivity of our actions. Or, if we work to change our current practices, perhaps future generations will applaud us. By promoting change, will we have secured a better future not only for animals but also for our ourselves and our grandchildren? We now look back on some of our own past public health practices and are chagrined at some of the very poor decisions we have made. It used to be common practice to advertise tobacco products in our most venerable medical journals. We withheld treatment for syphilis from impoverished African-Americans so that we could study the 'natural course' of the disease. During these times, less than 50 years ago, these practices were considered acceptable. Fortunately our thinking has evolved and continues to do so. We now recognize many of our past unethical or misguided practices for what they were.

We now have a choice before us. Do we use our knowledge to continue to condemn animals to incalculable harm, in turn jeopardizing our own health, or do we use that knowledge to evolve the practice of public health and improve the welfare of all? Do we continue to ignore the sad plight of animals who are abused, traded, eaten and used for experiments and consequently ignore how their plight affects our own health, or do we use our scientific advances and knowledge to fight against abuse, protect animals and their habitats, clothe ourselves without animal skins and fur, entertain ourselves without debasing animals, and feed ourselves and produce medicines without hurting animals?

We can do all of these things today. In fact, we are at amazing crossroads in human history. We can largely exist and, even more, exist better without compromising the welfare of animals. Curtailing our harmful practices against animals will significantly reduce a great many of the problems that currently threaten our health and welfare. How often in life are we given the opportunity to tackle several major obstacles to both our individual and collective health—and deal with the ethical conundrum of our poor treatment of animals—with rather simple solutions? In comparison with so many other obstacles that public health faces, such as poverty, war and social inequities, the improvement of animal welfare is often a relatively easy goal to accomplish. A gesture as simple as choosing one plate of food over another can single-handedly help thwart epidemics, curtail global warming and lengthen our lives—and reduce the number of animals in factory farms. By redirecting our medical resources toward the use and development of human-based tests, we can create far more predictive testing methods and avoid significant harm to animals. Striving to minimize the harms we cause to animals does not require us to abandon our quest to further human health. Rather, our endeavor to improve human health will be substantially advanced by promoting better treatment of animals.

All we need to do is acknowledge the harms we cause to animals and have the courage to move the public health field forward accordingly. We are starting to recognize our symbiotic relationship with animals—that our welfare is tied to theirs. Our treatment of animals is gaining more and more scrutiny and many people are advocating a change in their treatment. Public health just needs to embrace that change, help guide it and move it along at a faster pace. What is the legacy we wish to leave behind: one in which we ignore the plight of animals and jeopardize our own health or one in which we are at the forefront of recognizing the welfare of all who can suffer? It is time now for public health to continue its legacy of fighting for the underdog and in turn improve the health of all. It is time to include animals as part of the 'public' in public health.

Notes

1 The Welfare of Animals and Its Relevance to Our Health

1. Akhtar A. 'Flu farms?' *Science Progress* April 29, 2009. www.scienceprogress. org, date accessed November 19, 2010.
2. Fritz Institute. 'Fritz Institute-Harris interactive Katrina survey reveals inadequate immediate relief provided to those most vulnerable' April 26, 2006. www.fritzinstitute.org, date accessed July 1, 2010.
3. Hall MJ, Ng A, Ursano RJ, Holloway H, Fullerton C, Casper J. 'Psychological impact of the animal-human bond in disaster preparedness and response' *Journal of Psychiatric Practice* 2004; 10: 368–374.
4. Lowe SR, Rhodes JE, Zwiebach L, Chan CS. 'The impact of pet loss on the perceived social support and psychological distress of hurricane survivors' *Journal of Trauma Stress* 2009; 22: 244–247; Hunt M, Al-Awadi H, Johnson M. 'Psychological sequelae of pet loss following Hurricane Katrina' *Anthrozoos* 2008; 21: 109–121.
5. Fox Reno News. 'Tainted pet food causes spike in cat kidney failure' April 1, 2007. www.foxreno.com, date accessed August 16, 2011.
6. Food and Drug Administration. 'Melamine pet food recall of 2007' 2007. www.fda.gov, date accessed December 1, 2010.
7. Kluger J. 'Killer-whale tragedy: What made Tilikum snap?' *Time* February 26, 2010.
8. The Pew Charitable Trusts and Johns Hopkins Bloomberg School of Public Health. 'Putting meat on the table: Industrial farm animal production in America' *The Pew Commission on Industrial Farm Animal Production* 2008. www.ncifap.org, date accessed November 8, 2009; Food and Agriculture Organization of the United Nations. *Livestock's Long Shadow: Environmental Issues and Options* (Rome: Food and Agriculture Organization of the United Nations) 2006.
9. London RH. 'China kills 10 000 civet cats in "patriotic" campaign against SARS' *Student British Medical Journal* 2004; 12: 45–88.
10. Benatar D. 'The chickens come home to roost' *American Journal of Public Health* 2007; 97: 1545–1546.
11. Bateson P. 'Do animals feel pain?: By drawing analogies between humans and other animals, researchers tentatively conclude that fish and octopuses can feel pain, but insects can't. But the cutoff point is inevitably fuzzy' *New Scientist* April 25, 1992.
12. Sneddon LU. 'The evidence for pain in fish: The use of morphine as an analgesic' *Applied Animal Behaviour Science* 2003; 83: 153–162.
13. Akhtar S. 'Animal pain and welfare. Can pain sometimes be worse for them than for us?' pp. 495–518 in Beauchamp TL, Frey RJ (eds), The Oxford Handbook of Animal Ethics (USA: Oxford University Press) 2011.

14. Linzey A. *Why Animal Suffering Matters: Philosophy, Theology, and Practical Ethics* (New York: Oxford University Press) 2009, p. 10.
15. Balcombe J. *Pleasurable Kingdom: Animals and the Nature of Feeling Good* (New York: Palgrave Macmillan) 2006, p. 27; Balcombe J. *Second Nature: The Inner Lives of Animals* (New York: Palgrave Macmillan) 2010, p. 4.
16. Perry N. ' "Signing" chimp Washoe broke language barrier' *The Seattle Times* November 1, 2007.
17. Griffin DR. *Animal Minds: Beyond Cognition to Consciousness*, 2nd edn. (Chicago: University of Chicago Press) 1992, p. 239.
18. Smithsonian Zoological Park. 'Orangutan language project, think tank research projects' http://nationalzoo.si.edu, date accessed September 1, 2010.
19. Griffin, 1992, pp. 237, 265.
20. Whitten A. 'From the field to the laboratory and back again: Culture and "social mind" in primates' in Bekoff M, Allen C, Gordon M (eds), *The Cognitive Animal: Empirical and Theoretical Perspectives on Animal Cognition* (Cambridge: MIT Press) 2002; Balcombe, 2010, p. 127.
21. Tomasello M, Call J, Hare B. 'Chimpanzees understand psychological states— the question is which ones and to what extent' *Trends in Cognitive Science* 2003; 7: 153–156; Hare B, Call J, Agnetta B, Tomasello M. 'Chimpanzees know what conspecifics do and do not see' *Animal Behavior* 2000; 59: 771–785; Hare B, Call J, Tomasello M. 'Do chimpanzees know what conspecifics know?' *Animal Behaviour* 2001; 63: 139–151.
22. Pokorny JJ, de Waal FBM. 'Monkeys recognize the faces of group mates in photographs' *Proceedings of the National Academy of Sciences* 2009; 106: 21539–21543.
23. Hampton RR. 'Rhesus monkeys know when they remember' *Proceedings of the National Academy of Sciences* 2001; 98: 5359–5362; Hampton RR, Hampstead BM. 'Spontaneous behavior of a rhesus monkey (*Macaca mulatta*) during memory tests suggests memory awareness' *Behavioral Processes* 2006; 72: 184–189.
24. Hampton RR, 2001.
25. Church RM. 'Emotional reactions of rats to the pain of others' *Journal of Comparative Physiological Psychology* 1959; 52: 132–134.
26. Langford DJ, Crager SE, Shehzad Z et al. 'Social modulation of pain as evidence for empathy in mice' *Science* 2006; 312: 1967–1970.
27. de Waal FBM. 'Commiserating mice' *Scientific American News Blog* July 24, 2007. www.scientificamerican.com, date accessed December 10, 2010.
28. Panksepp J. 'Beyond a joke: From animal laughter to human joy?' *Science* 2005; 308: 62–63.
29. BBC News. 'Moral maze. Model behavior' 2010. http://www.bbc.co.uk, date accessed September 14, 2010; Wilkinson GS. 'Reciprocal food sharing in the vampire bat' *Nature* 1984; 308: 181–184.
30. Highfield R. 'So who are you calling bird brain? Chatter of chickens proves they are brighter than we thought' *Telegraph* November 15, 2006.
31. Evans CS, Evans L. 'Representational signaling in birds' *Biology Letters* 2007; 3: 8–11.
32. Grimes W. 'If chickens are so smart. Why aren't they eating us?' *New York Times* January 12, 2003.

33. Vallortigara G, Regolin L, Rigoni M, Zanforlin M. 'Delayed search for a concealed imprinted object in the domestic chick' *Animal Cognition* 1998; 1: 17–24; Regolin L, Vallortigara G, Zanforlin M. 'Object and spatial representations in detour problems by chicks' *Animal Behaviour* 1995; 49: 195–199.
34. Emery NJ, Clayton NS. 'Effects of experience and social context on prospective caching strategies by scrub jays' *Nature* 2001; 414: 443–446; Clayton NS, Dickinson A. 'Episodic-like memory during cache recovery by scrub jays' *Nature* 1998; 395: 272–274.
35. *ScienceDaily*. 'Can you ask a pig if his glass is half full?' July 28, 2010. www.sciencedaily.com, date accessed September 14, 2010.
36. Broom DM, Sena H, Moynihan KL. 'Pigs learn what a mirror image represents and use it to obtain information' *Animal Behaviour* 2009; 78: 1037–1041.
37. Balcombe, 2010, p. 48.
38. de Waal FBM. 'Darwin's last laugh' *Nature* 2009; 460: 175.
39. Baars BJ. *In the Theatre of Consciousness: The Workspace of the Mind* (New York: Oxford University Press) 1997, pp. 33–34.
40. Bekoff M. 'Animal passions and beastly virtues: Cognitive ethology as the unifying science for understanding the subjective, emotional, empathic, and moral lives of animals' *Zygon* 2006; 41: 71–104.
41. de Waal, 2009.
42. Balcombe, 2010, pp. 46–47.
43. Rollin BE. 'Animal pain' in Armstrong SJ, Botzler RG (eds), *The Animal Ethics Reader*, 2nd edn. (New York: Routledge) 2008, p. 136.
44. US Department of Health and Human Services, Office of Laboratory Animal Welfare. 'Public health service policy on humane care and use of laboratory animals' 2002. http://grants.nih.gov/grants/olaw/references/phspol.htm?, date accessed August 8, 2011.
45. Bekoff M. 'Animal reflections' *Nature* 2002; 419: 255.
46. Goldsmith TH. 'Hummingbirds see near ultraviolet light' *Science* 1980; 207: 786–788; Ventura DF, Zana Y, de Souza JM, DeVoe RD. 'Ultraviolet colour opponency in the turtle retina' *Journal of Experimental Biology* 2001; 204: 2527–2534; Varela FJ, Thompson E. 'Color vision: A case study in the foundations of cognitive science' Revue de Synthèse 1990; 111: 129–138.
47. Ibid.; Emmerton J, Delhis JD. 'Wavelength discrimination in the "visible" and ultraviolet spectrum by pigeons' *Journal of Comparative Physiology* 1980; 141: 47–52.
48. Goldsmith, 1980.
49. Rollin BE. 'Animal research: A moral science' *EMBO Reports* 2007; 8: 521–525.
50. Akhtar, 2010.
51. Rollin, 2007.
52. Bekoff, 2002.
53. Johnson, S. *Ghost Map: The Story of London's Most Terrifying Epidemic—and How It Changed Science, Cities and the Modern World* (New York: Riverhead Books) 2006; Newsom SWB. 'Pioneers in infection control: John Snow, Henry Whitehead, the Broad Street pump, and the beginnings of geographical epidemiology' *Journal of Hospital Infection* 2006; 64: 210–216.
54. Johnson, 2006; Newsom, 2006.

55. Ramsay MAE. 'John Snow, MD: Anaesthetist to the Queen of England and pioneer epidemiologist' *Proceedings (Baylor University Medical Center)* 2006; 19: 24–28.
56. Irwin A, Scali E. 'Action on the social determinants of health: A historical perspective' *Global Public Health* 2007; 2: 235–256.
57. Ibid.
58. Ruffin J. 'The science of eliminating health disparities: Embracing a new paradigm' *American Journal of Public Health* 2010; 100: S8–S9.
59. Donohoe M. 'Roles and responsibilities of health care professionals in combating environmental degradation and social injustice: Education and activism' *Monash Bioethics Review* 2008; 27: 65–82.
60. *New Internationalist* magazine. 'A history of public health' 2001. www. thirdworldtraveler.com, date accessed October 26, 2009.
61. Terris M. 'Evolution of public health and preventive medicine in the United States' *American Journal of Public Health* 1975; 65: 161–169.
62. Winkelstein W. 'History of public health' *Encyclopedia of Public Health* 2009. http://www.enotes.com, date accessed September 29, 2009.
63. Centers for Disease Control and Prevention. 'Ten great public health achievements—United States, 1900–1999' *Morbidity and Mortality Weekly Report* 1999; 48: 241–243.
64. Wardell D. 'Margaret Sanger: Birth control's successful revolutionary' *American Journal of Public Health* 1980; 70: 736–742.
65. Ibid.
66. Ibid.
67. Sadler, JS. 'Commentary: Stigma, conscience, and science in psychiatry: Past, present, and future' *Academic Medicine* 2009; 84: 413–417.
68. World Health Organization. 'Preamble to the Constitution of the World Health Organization', as adopted by the International Health Conference, New York, June 19–July 22, 1946; signed on July 22, 1946 by the representatives of 61 States. Off. Rec. Wld. Hlth. Org. 2, 100' 1946.
69. Irwin and Scalia, 2007.
70. Coleman C, Bouesseau MC, Reis A. 'The contribution of ethics to public health' *Bulletin of the World Health Organization* 2008; 86: 578; Arras JD, London AJ (eds), *Ethical Issues in Modern Medicine*, 6th edn. (New York: McGraw-Hill) 2003, pp. 693–703.
71. Gruskin S, Mills EJ, Tarantola D. 'History, principles, and practice of health and human rights' *Lancet* 2007; 370: 449–458.
72. Ibid.
73. Marks SP. 'Jonathan Mann's legacy to the 21st century: The human rights imperative for public health' *Journal of Law, Medicine & Ethics* 2001; 29: 131–138.
74. Mann JM. 'Health and human rights: If not now, when?' *Health and Human Rights* 1997; 2: 113–120.
75. Foege, WH. 'Global public health: Targeting inequities' *Journal of the American Medical Association* 1998; 279: 1931–1932.
76. Tarlov A. 'Public policy frameworks for improving population health' *Annals of the New York Academy of Sciences* 1999; 896: 281–293.
77. Leeder S. 'The scope, mission, and method of contemporary public health' *Australian and New Zealand Journal of Public Health* 2007; 31: 505–508.

78. Board on Health Promotion and Disease Prevention and Institute of Medicine. *The Future of the Public's Health in the 21st Century, 2002* (Washington, DC: The National Academies Press) 2002, pp. 38–54; Winkelstein, 2009; McMichael A, Butler C. 'Emerging health issues: The widening challenge for population health promotion' *Health Promotion International* 2007; 21: 15–24.
79. Food and Agriculture Organization of the United States. *FAOSTAT*, 2009. http://faostat.fao.org, date accessed December 4, 2010.
80. Worldwatch Institute. *State of the World 2006* 2006, p. 26. http://www.worldwatch.org, date accessed December 10, 2010.
81. Taylor K, Gordon N, Langley G et al. 'Estimates for worldwide animal use in 2005' *ATLA* 2008; 36: 327–342.
82. American Pet Products. 'Industry statistics and trends, 2009/2010' 2010. www.americanpetproducts.org, date accessed August 29, 2010.
83. Brady D, Palmeri C. 'The pet economy' *Bloomberg Business Week* August 6, 2007.
84. Ibid.
85. BBC News. 'EU bans battery hen cages' January 28, 1999. http://news.bbc.co.uk, date accessed September 21, 2008.
86. 'China unveils first ever animal cruelty legislation' *Daily Telegraph* September 18, 2009. http://blogs.telegraph.co.uk, date accessed September 18, 2009.
87. *New Scientist.* 'Sentient beings' June 28, 1997. http://www.newscientist.com, date accessed December 12, 2010.
88. Balcombe, 2010, p. 195.
89. Ibid.
90. Carter T. 'Beast practices' *ABA Journal* 2007. www.abajournal.com, date accessed September 19, 2010.
91. The British Sociological Association. 'Animal/Human Study Groups'. www.britsoc.co.uk, date accessed October 10, 2010.

2   Victims of Abuse: Making the Connection

1. Berg M. '"Hapless dependents": Women and animals in Anne Bronte's *Agnes Grey*' *Studies in the Novel* Summer 2002.
2. Flynn CP. 'Acknowledging the zoological connection: A sociological analysis of animal cruelty' *Society & Animals* 2001; 9: 71–87.
3. Ibid.
4. Zilney LA, Zilney M. 'Reunification of child and animal welfare agencies: Cross-reporting of abuse in Wellington county, Ontario' *Child Welfare* 2005; LXXXIV (1): 47–66; BBC News. 'Children as animals—origins of anti-cruelty laws' December 5, 2001. www.bbc.co.uk, date accessed November 11, 2009.
5. BBC News, 2001; Zilney and Zilney, 2005.
6. Zilney and Zilney, 2005.
7. Radford University. 'Environmental History Timeline'. www.radford.edu, date accessed February 14, 2011.

8. Goleman D. 'Clues to a dark nurturing ground for one serial killer' *New York Times* August 7, 1991.
9. Albert B. 'Experts say animal abuse can be a warning sign' *Milford-Orange Bulletin* July 28, 2010.
10. Lockwood R, Hodge GH. 'The tangled web of animal abuse: The links between cruelty to animals and human violence' in Lockwood R, Ascione FR (eds), *Cruelty to Animals and Interpersonal Violence* (West Lafayette, IN: Purdue University Press) 1998, pp. 77–82; Schecter H, Everitt D. *The A to Z Encyclopedia of Serial Killers* (New York: Pocket Books) 1996.
11. Potter T. 'BTK describes his own crimes' *The Wichita Eagle* July 16, 2005.
12. Bradley P, Krishnamurthy K. 'Right and wrong "an illusion"/psychologist who met with Malvo Said teen's disorder limited his moral judgment' *Richmond Times Dispatch* December 9, 2003.
13. Schecter and Everitt, 1996.
14. Lockwood and Hodge, 1998.
15. Peterson ML and Farrington DP(a). 'Measuring animal cruelty and case histories' in Linzey A (ed.), *The Link between Animal Abuse and Human Violence* (Portland, OR: Sussex Academic Press) 2009, pp. 3–23.
16. Ibid.
17. Neustatter A. 'Killers' animal instincts: The sadistic fantasies that drive serial killers have their roots in childhood—There is a compelling link with cruelty to animals' *Independent* October 13, 1998; Ascione FR. 'The abuse of animals and human interpersonal violence: Making the connection' in Ascione FR, Arkows P (eds), *Child Abuse, Domestic Violence, and Animal Abuse: Linking the Circles of Compassion for Prevention and Intervention* (West Lafayette, IN: Purdue University Press) 1999, pp. 50–61.
18. Zuckoff M. 'Loners Drew Little Notice' *Boston Globe* April 22, 1999.
19. Peterson and Farrington (a), 2009.
20. Bronner E. 'Terror in Littleton: The signs; experts urge swift action to fight depression, isolation, and aggression' *New York Times* April 22, 1999.
21. Wilson P, Norris G. 'Relationship between criminal behavior and mental illness in young adults: Conduct disorder, cruelty to animals and young adult serious violence' *Psychiatry, Psychology and Law* 2003; 10: 239–243; Peter and Farrington (a), 2009.
22. Schecter and Everitt, 1996.
23. Peterson and Farrington (a), 2009.
24. Goleman, 1991.
25. Peterson and Farrington (a), 2009.
26. Ibid.; Flynn, 2001.
27. Quinn KM. 'Violent behavior. Animal abuse at early age linked to interpersonal violence' *The Brown University Child and Adolescent Behavior Letter* March 2000; 16 (3).
28. Miller KS, Knutson JF. 'Reports of severe physical punishment and exposure to animal cruelty by inmates convicted of felonies and by university students' *Child Abuse and Neglect* 1997; 21: 59–82; Tallichet SE, Hensley C, Singer SD (a). 'Unraveling the methods of childhood and adolescent cruelty to nonhuman animals' *Society and Animals* 2005; 13: 91–106.
29. Flynn, 2001.

30. Tallichet SE, Hensley C, O'Bryan A, Hassal H (b). 'Targets for cruelty: Demographic and situational factors affecting the type of animal abused' *Criminal Justice Studies* 2005; 18: 173–182.
31. Hensley C, Tallichet SE. 'The effect of inmates' self-reported childhood and adolescent animal cruelty—Motivations on the number of convictions for adult violent inter-personal crimes' *International Journal of Offender Therapy and Comparative Criminology* 2008; 52: 175–184.
32. Merz-Perez L, Heide K, Silverman I. 'Childhood cruelty to animals and subsequent violence against humans' *International Journal of Offender Therapy and Comparative Criminology* 2001; 45: 556–573.
33. Schiff K, Louwe D, Ascione FR. 'Animal relations in childhood and later violent behavior against humans' *Acta Criminologica* 1999; 12: 76–77.
34. Kellert SR, Felthous AR. 'Childhood cruelty toward animals among criminals and noncriminals' *Human Relations* 1985; 38: 1113–1129.
35. Tingle D, Barnard GW, Robbins G, Newman G, Hutchinson D. 'Childhood and adolescent characteristics of pedophiles and rapists' *International Journal of Law and Psychiatry* 1986; 9: 103–116.
36. Johnson BR, Becker JV. 'Natural born killers? The development of a sexually sadistic serial killer' *Journal of the American Academy of Psychiatry and the Law* 1997; 25: 335–348.
37. Ressler K, Burgess AW, Douglas JE. *Sexual Homicide: Patterns and Motives* (Lexington, MA: Lexington Books) 1988.
38. Hensley C, Tallichet SE, Singer SD. 'Exploring the possible link between childhood and adolescent bestiality and interpersonal violence' *Journal of Interpersonal Violence* 2006; 21: 910–923.
39. Simons DA, Wurtle SK, Durham RL. 'Developmental experiences of child sexual abusers and rapists' *Child Abuse & Neglect* 2008; 32: 549–560.
40. Humane Society of the United States (HSUS). 'Animal cruelty, serial killers, and violence'. www.hsus.org, date accessed July 2, 2010.
41. Boat BW, Knight JC. 'Experiences and needs of adult protective services case managers when assisting clients who have companion animals' *Journal of Elder Abuse & Neglect* 2000; 12: 145–155.
42. HSUS. 'Animal cruelty, serial killers, and violence'.
43. Arluke A, Levin J, Luke C, Ascione FR. 'The relationship of animal abuse to violence and other forms of antisocial behavior' *Journal of Interpersonal Violence* 1999; 14: 963–975; Luke C, Arluke A, Levin J. 'Cruelty to animals and other crimes. A study by the MSPCA and Northeastern University, December 1997'. www.mspca.org, date accessed December 10, 2010.
44. Brown C. 'Dogfighting charges filed against Falcons' Vick' *New York Times* July 18, 2007.
45. Ibid.; Maske M. 'Falcons' Vick indicted in dogfighting case' *Washington Post* July 18, 2007.
46. Brown, 2007.
47. Simmons R. 'Dog eat dog: The bloodthirsty underworld of dogfighting' *The OC Dog Newspaper* December 26, 2005. www.ocdogs.com, date accessed December 13, 2010.
48. Gleyzer R, Felthous AR, Holzer CE. 'Animal cruelty and psychiatric disorders' *Journal of the American Academy of Psychiatry and the Law* 2002; 30: 257–265.

49. Vaughn MG, Fu Q, DeLisi M et al. 'Correlates of cruelty to animals in the United States: Results from the national epidemiologic survey on alcohol and related conditions' *Journal of Psychiatric Research* 2009; 43: 1213–1218.
50. Gordon MS, Kinlock TW, Battjes RJ. 'Correlates of early substance abuse and crime among adolescents entering outpatient substance abuse treatment' *American Journal of Drug and Alcohol Abuse* 2004; 30: 39–59.
51. Shapiro HL, Prince JB, Ireland R, Stein MT 'A dominating imaginary friend, cruelty to animals, social withdrawal, and growth deficiency in a 7-year-old girl with parents with schizophrenia' *Journal of Developmental and Behavioral Pediatrics* 2010; 31: S24–S29; Sherley M. 'Why doctors should care about animal cruelty' *Australian Family Physician* 2007; 36: 61–63; Tucker HS, Finlay F, Guiton S. 'Munchausen syndrome involving pets by proxies' *Archives of Disease in Childhood* 2002; 87: 263.
52. Degrue S, DiLillo D. 'Is animal cruelty a "red flag" for family violence? Investigating co-occurring violence toward children, partners, and pets' *Journal of Interpersonal Violence* 2009; 24: 1036–1056.
53. Walsh F. 'Human-animal bonds II. The role of pets in family systems and family therapy' *Fam Proc* 2009; 48: 481–499.
54. DeGrue and DiLillo, 2009.
55. 'UNCC research links animal abuse, crimes against people' *Community Policing Digest* 1998; 3: 3–4.
56. Siebert C. 'The animal-cruelty syndrome' *New York Times* June 7, 2010.
57. Ibid.
58. Ibid.
59. Gupta M. 'Functional links between intimate partner violence and animal abuse: Personality features and representations of aggression' *Society & Animals* 2008; 16: 223–242; Gallagher B, Allen M, Jones B. 'Animal abuse and intimate partner violence: Researching the link and its significance in Ireland—A veterinary perspective' *Irish Veterinary Journal* 2008; 61: 658–667.
60. Flynn CP. 'Women's best friend: Pet abuse and the role of companion animals in the lives of battered women' *Violence against Women* 2000; 6: 162–177.
61. Marcus C, Kent M. 'Abused women risk their lives for pets: Study' *Sunday Age* March 9, 2008.
62. Ascione FR. 'Battered women's reports of their partners' and their children's cruelty to animals' *Journal of Emotional Abuse* 1998; 1: 119–133.
63. Quinlisk JA. 'Animal abuse and family violence' in Ascione FR, Arkow P (eds), *Child Abuse, Domestic Violence, and Animal Abuse: Linking the Circles of Compassion for Prevention and Intervention* (West Lafayette, IN: Purdue Univeristy Press) 1999, pp. 168–175.
64. Siebert, 2010.
65. Volant AM, Johnson JA, Gullone E, Coleman GJ. 'The relationship between domestic violence and animal abuse: An Australian study' *Journal of Interpersonal Violence* 2008; 23: 1277–1295.
66. Ascione FR, Weber CV, Thompson TM et al. 'Battered pets and domestic violence: Animal abuse reported by women experiencing intimate partner violence and by nonabused women' *Violence against Women* 2007; 13: 354–373.
67. Siebert, 2010.

68. Simmons CA, Lehmann P. 'Exploring the link between pet abuse and controlling behaviors in violent relationships' *Journal of Interpersonal Violence* 2007; 22: 1211–1222.
69. Golson J. 'Awareness of pets victimized in domestic abuse grows' *The Star Ledger* May 30, 2010.
70. Quinn, 2000.
71. Flynn CP. 'Exploring the link between corporal punishment and children's cruelty to animals' *Journal of Marriage and Family* 1999; 61: 971–981.
72. Ross S. 'Abuse victims who refuse to leave for fear of their pets being harmed' *The Scotsman* October 20, 2008.
73. Faver CA, Strand EB. 'To leave or stay? Battered women's concern for vulnerable pets' *Journal of Interpersonal Violence* 2003; 18: 1367–1377.
74. Ascione FR. 'Battered women's reports of their partners and their children's cruelty to animals' *Journal of Emotional Abuse* 1998; 1: 199–233.
75. Ross, 2008.
76. Gallagher et al., 2008; Marcus C. 'Victims of domestic violence; women suffer for sake of their pets' *The Sun Herald* March 9, 2008; Pollet, SL. 'Outside counsel; news; the link between animal abuse and family violence' *New York Law Journal* 2008; 239: 14; Agrell S. 'Pets help abused women—But may make them vulnerable' *The Globe and Mail* June 12, 2008.
77. Ibid.; Pollet, 2008; Ross, 2008; *Washington Post*. 'Sheltering women—And their pets, too; experience with victims who stayed in violent homes for the sake of their dog or cat led a humane association official to help make room for animals at Arlington shelter' *Washington Post* November 8, 2007.
78. Marcus and Kent, 2008.
79. Agrell, 2008; Peters S. 'A haven for abuse victims who keep their pets close; organizations push shelters to offer on-site boarding' *USA Today* March 18, 2008; *Washington Post* 2007; Ross, 2008.
80. Peters, 2008.
81. Ross, 2008.
82. Peters, 2008.
83. Ross, 2008.
84. Peters, 2008.
85. Ibid.; *Washington Post* 2007.
86. Peters, 2008.
87. American Human Association. 'American Humane Association recognizes Paula Abdul for her support of the Pets and Women's Shelters (PAWS)® program' October 7, 2010. www.americanhumane.org, date accessed February 11, 2011.
88. Paws for Kids. www.pawsforkids.org.uk, date accessed July 21, 2011.
89. *Washington Post*, 2007; Peters, 2008.
90. Kogan, LR, McConnell S, Schoenfeld-Tacher R, Jansen-Lock P. 'Crosstrails: A unique foster program to provide safety for pets of women in safehouses' *Violence against Women* 2004; 10: 418–434.
91. Gallagher et al., 2008.
92. Ross, 2008.
93. Tapia F. 'Children who are cruel to animals' *Child Psychiatry and Human Development* 1971; 2: 70–77.

94. Felthous AR. 'Aggression against cats, dogs, and people' *Child Psychiatry and Human Development* 1980; 10: 169–177.
95. Flynn, 1999.
96. Currie CL. 'Animal cruelty by children exposed to domestic violence' *Child Abuse & Neglect* 2006; 30: 425–435.
97. Becker KD, Stuewig J, Herrera VM, McCloskey LA. 'A study of firesetting and animal cruelty in children: Family influences and adolescent outcomes' *Journal of the American Academy of Child and Adolescent Psychiatry* 2004; 43: 905–912.
98. Ascione FR, Friedrich WN, Heath J, Hayashi K. 'Cruelty to animals in normative, sexually abused, and outpatient psychiatric samples of 6- to 12-year old children: Relations to maltreatment and exposure to domestic violence' *Anthrozoos* 2003; 16: 194–212.
99. DeViney E, Dickert J, Lockwood R. 'The care of pets within child abusing families' *International Journal for the Study of Animal Problems* 1983; 4: 321–329.
100. Hutton JS. 'Animal abuse as a diagnostic approach in social work: A pilot study' in Lockwood R, Ascione FR (eds), *Cruelty to Animals and Interpersonal Violence* (West Lafayette, IN: Purdue University Press) 1998, pp. 415–418.
101. Duncan A, Thomas JC, Miller C. 'Significance of family risk factors in development of childhood animal cruelty in adolescent boys with conduct problems' *Journal of Family Violence* 2005; 20: 235–239.
102. Henry BC. 'Empathy, home environment, and attitudes toward animals in relation to animal abuse' *Anthrozoos* 2006; 19: 17–34.
103. Sherley, 2007.
104. Vollum S, Buffington-Vollum J, Longmire DR. 'Moral disengagement and attitudes about violence toward animals' *Society & Animals* 2004; 12: 209–235.
105. Quinlisk, 1999.
106. Duncan A, Miller C. 'The impact of an abusive family context on childhood animal cruelty and adult violence' *Aggression and Violent Behavior* 2002; 7: 365–383.
107. Quinn, 2000.
108. Flynn, 2001.
109. Loar L. ' "I'll only help you if you have two legs," or, why human services professionals should pay attention to cases involving cruelty to animals' in Ascione FR, Arkow P (eds), *Child Abuse, Domestic Violence, and Animal Abuse: Linking the Circles of Compassion for Prevention and Intervention* (West Lafayette, IN: Purdue University Press) 1999, pp. 120–136; Ascione FR. *Children and Animals: Exploring the Roots of Kindness and Cruelty* (West Lafayette, IN: Purdue University Press) 2005.
110. Sherley, 2007.
111. Quinn, 2000.
112. Robin M, Ten Bensel R, Quigley J. 'A study of the relationship of childhood pet animals and the psychosocial development of adolescents'. Paper presented at the International Conference on the Human/Companion Animal Bond, Philadelphia, PA, 1981.

113. Daly B, Morton LL. 'Empathic correlates of witnessing the inhumane killing of an animal: An investigation of single and multiple exposures' *Society & Animals* 2008; 16: 243–255.
114. Hensley C, Tallichet SE. 'Learning to be cruel? Exploring the onset and frequency of animal cruelty' *International Journal of Offender Therapy and Comparative Criminology* 2005; 49: 37–47.
115. Flynn, 2001.
116. Henry BC, Sanders CE. 'Bullying and animal abuse: Is there a connection?' *Society & Animals* 2007; 15: 107–126.
117. Verlinden S, Herson M, Thomas J. 'Risk factors in school shootings' *Clinical Psychology Review* 2000; 29: 3–56.
118. Baldry AC. 'Animal abuse among preadolescents directly and indirectly victimized at school and at home' *Criminal Behavior and Mental Health* 2005; 15: 97–110.
119. Gullone E, Robertson N. 'The relationship between bullying and animal abuse behaviors in adolescents: The importance of witnessing animal abuse' *Journal of Applied Developmental Psychology* 2008; 29: 371–379.
120. Henry and Sanders, 2007.
121. Hackett S, Uprichard E. 'Animal abuse and child maltreatment: A review of the literature and findings from a UK study' NSPCC 2007. www.nspcc.org.uk, date accessed February 10, 2011.
122. Beirne P. 'From animal abuse to interhuman violence? A critical review of the progression thesis' *Society & Animals* 2004; 12: 39–65.
123. Peterson ML, Farrington DP(b). 'Types of cruelty: Animals and childhood cruelty, domestic violence, child and elder abuse' in Linzey (ed.), *The Link between Animal Abuse and Human Violence* (Portland, OR: Sussex Academic Press) 2009, pp. 24–37.
124. Ibid.
125. McPhedran S. 'A Review of the evidence for associations between empathy, violence, and animal cruelty' *Aggression and Violent Behavior* 2009; 14: 1–4; Flynn, 2001.
126. Arluke et al., 1999; Hackett and Uprichard, 2007.
127. Arluke et al., 1999.
128. Levin J, Arluke A. 'Reducing the link's false positive problem' in Linzey A (ed.), The *Link between Animal Abuse and Human Violence* (Sussex Academic Press: Eastbourne, UK) 2009, pp. 164–171.
129. Peterson and Farrington (b), 2009.
130. Felthous AR, Kellert SR. 'Childhood cruelty to animals and later aggression against people: A review' *American Journal of Psychiatry* 1987; 144: 710–717.
131. Tallichet et al. (a), 2005.
132. Ascione F. 'Animal abuse and youth violence' *OJJDP Juvenile Justice Bulletin* September 2001. www.ncjrs.org, date accessed February 10, 2011.
133. Tallichet et al. (b) 2005; Beirne 2004; Flynn, 2001.
134. Vaca-Guzman M, Arluke A. 'Normalizing passive cruelty: The excuses and justifications of animal hoarders' *Anthrozoos* 2005; 18: 338–357.
135. Becker F. 'The links between child abuse and animal abuse' *National Society for the Prevention of Cruelty to Children* 2001. www.nspcc.org.uk, date accessed November 14, 2009.

136. Felthous and Kellert, 1987; Currie CL. 'Animal cruelty by children exposed to domestic violence' *Child Abuse & Neglect* 2006; 30: 425–435; Baldry AC. 'The development of the P.E.T. scale for the measurement of physical and emotional tormenting against animals in adolescence' *Society & Animals* 2004; 12: 1–17.

137. Offord DR, Boyle MC, Racine YA. 'The epidemiology of antisocial behavior in childhood and adolescence' in Pepler DJ, Rubin KH (eds), *The Development and Treatment of Childhood Aggression* (Hillsdale, NJ: Lawrence Erlbaum Associates) 1991, pp. 31–54.

138. Ascione, 1999; Boat, BW. 'The relationship between violence to children and violence to animals. An ignored link?' *Journal of Interpersonal Violence* 1995; 10: 229–235.

139. Beirne, 2004.

140. Ibid.; Taylor N, Signal TD. 'Community demographics and the propensity to report animal cruelty' *Journal of Applied Animal Welfare Science* 2006; 9: 201–210.

141. Beirne, 2004.

142. Ibid.

143. Baldry AC. 'Animal abuse and exposure to interparental violence in Italian youth' *Journal of Interpersonal Violence* 2003; 18: 258–281; Flynn, 2001.

144. Becker, 2001.

145. Arluke et al., 1999.

146. Becker, 2001.

147. Duncan A, Miller C. 'The impact of an abusive family context on childhood animal cruelty and adult violence' *Aggression and Violent Behavior* 2002; 7: 365–383.

148. Ibid.

149. Barnes R. 'Supreme Court overturns anti-animal cruelty law in First Amendment case' *Washington Post* April 21, 2010.

150. Linzey A (ed.). 'Does animal abuse really benefit us?', *The Link between Animal Abuse and Human Violence* (Sussex: Academic Press) 2009, pp. 1–10.

151. Blewitt J. 'The link between animal cruelty and child protection' *Community Care* October 30, 2008.

152. Sherley, 2007.

153. Currie, 2006; Flynn, 2001.

154. Becker, 2001.

155. Flynn, 2001.

156. Raupp CD. 'The "furry ceiling". Clinical psychology and human–animal studies' *Society & Animals* 2002; 10: 353–360; Faver CA, Strand EB. 'Domestic violence and animal cruelty: Untangling the web of abuse' *Journal of Social Work Education* 2003; 39: 237–253.

157. Raupp, 2002.

158. Zilney and Zilney, 2005.

159. Hackett and Uprichard, 2007.

160. Blewitt, 2008; Faver and Strand, 2003; Urbina I. 'Animal abuse as clues to additional cruelties' *New York Times* March 17, 2010.

161. Blewitt, 2008.

162. Becker, 2001.

163. Zilney and Zilney, 2005.

164. Urbina, 2010; Pollet, 2008.
165. Beetz AM. 'Empathy as an indicator of emotional development' in Linzey A (ed.), *The Link between Animal Abuse and Human Violence* (Sussex: Academic Press) 2009, pp. 62–74.
166. Poresky RH. 'The young children's empathy measure: Reliability, validity, and effects of companion animal bonding' *Psychological Reports* 1990; 66: 931–936.
167. Thompson KL, Gullone E. 'Promotion of empathy through prosocial behavior in children through humane education' *Australian Psychologist* 2003; 38: 175–182.
168. Hein GE. 'Massachusetts Society for the Prevention of Cruelty to Animals outreach program evaluation: Final report' (Boston, MA: SPCA) 1987; Beetz, 2009; Thompson KL, Gullone E. 'Promotion of empathy through prosocial behavior in children through humane education' *Australian Psychologist* 2003; 38: 175–182; Ascione FR. 'Enhancing childrens' attitude about the human treatment of animals: Generalizations to human directed empathy' *Anthrozoos* 1992; 5: 176–191.
169. Flynn, 2001.
170. Arluke A, Luke C. 'Physical cruelty towards animals in Massachussetts, 1975–1996' *Society & Animals* 1997; 5: 195–201.
171. McKinley J. 'Lawmakers consider an animal abuse registry' *New York Times* February 21, 2010.
172. Urbina, 2010.
173. Mack SK. 'Law protects pets of abuse victims' *Bangor Daily News* April 1, 2006; Pollet, 2008.
174. Arkow P. 'Expanding domestic violence protective orders to include companion animals' *American Bar Association Commission on Domestic Violence eNewsletter* 2007; 8.
175. Sherley, 2007.

3 Lions, Tigers and Bears: The Global Trade in Animals

1. Farinato R. 'USDA seizes the moment, orders Hawthorne to give up 16 elephants' *Humane Society of the United States*, March 25, 2004. www.hsus.org, date accessed July 25, 2010; Sabatini M. 'Trainer tells of killer elephant's history of trouble, bad attitude' *Los Angeles Times* August 27, 1994.
2. Farinato, 2004.
3. Ibid.
4. The Elephant Sanctuary. 'In memory of Lota' www.elephants.com/lota/lotaBio.php, date accessed June 23, 2010.
5. Farinato, 2004; Elephant Sanctuary, 'In memory ...'.
6. Elephant Sanctuary, 'In memory ...'.
7. Ibid.
8. Bradshaw GA, Schore AN, Brown JL, et al. 'Elephant breakdown. Social trauma: early disruption of attachment can affect the physiology, behavior and culture of animals and humans over generations' *Nature* 2005; 433: 807.

9. Humane Society of the United States (HSUS). 'Wildlife trade'. www.hsus. org, date accessed March 15, 2010.

10. Alacs EA, Georges A, FitzsSimmons NN, Robertson J. 'DNA detective: A review of molecular approaches to wildlife forensics' *Forensics Science, Medicine and Pathology* December 16, 2009 (Epub ahead of print); Swift L, Hunter PR, Lees AC, Bell DJ. 'Wildlife trade and the emergence of infectious diseases' *Ecohealth* 2007; 4: 25–30; Sullivan M. 'Southeast Asia's illegal wildlife trade. Profits, demand fuel commerce in endangered species' *National Public Radio* November 3–5, 2003. www.npr.org/programs/re/archivesdate/2003/nov/wildlife/index.html, date accessed May 12, 2010.

11. Bilger B. 'Swamp things: Florida's uninvited predators' *The New Yorker* April 20, 2009: 80; Bushmeat Crisis Taskforce. 'Asian wildlife trade'. www. bushmeat.org, date accessed June 14, 2010; Baker SC. 'A truer measure of the market: The molecular ecology of fisheries and wildlife trade' *Molecular Ecology* 2008; 17: 3985–3998; Karesh WB, Cook RA, Gilbert M, Newcomb J. 'Implication of wildlife trade on the movement of avian influenza and other infectious diseases' *Journal of Wildlife Diseases* 2007; 43: S55–S59.

12. Aldhous P. 'Exotic pets pose risks to native species' *New Scientist* August 1, 2007. www.newscientist.com, date accessed June 21, 2010; Derr M. 'Lure of the exotic stirs trouble in the animal kingdom' *New York Times* February 12, 2002.

13. Aldhous, 2007.

14. Pavlin BI, Schloegel LM, Daszak P. 'Risks of importing zoonotic diseases through wildlife trade, United States' *Emerging Infectious Diseases* 2009; 15: 1721–1726.

15. Derr, 2002.

16. Humane Society of the United States (HSUS). 'The trade in live reptiles: Exports from the United States'. www.hsus,org, date accessed April 20, 2010.

17. Bergman C. 'Stolen animals are a $10 billion business worldwide. A journey into the Amazon Basin traces the illicit origins of wildlife trafficking' *Smithsonian* 2009; 40: 34–41; Adam D. 'Illegal pet trade is emptying the world's forests' *Guardian* (UK) February 25, 2010; Humane Society of the United States (HSUS). 'Wildlife trade'. www.hsus.org, date accessed March 15, 2010; Sullivan, 2003; Carpenter AI, Rowcliffe JM, Watkinson AR. 'The dynamics of the global trade in chameleons' *Biological Conservation* 2004; 120: 291–301.

18. Zhou Z, Jiang Z. 'Characteristics and risk assessment of international trade in tortoises and freshwater turtles in China' *Chelonian Conservation and Biology* 2008; 7: 28–36.

19. Kaur M, Aziz S. 'Malaysians "fund" illegal exotic wildlife trade' *New Straits Times* (Malaysia) October 6, 2006; Sullivan, 2003; Bergman, 2009.

20. Bergman, 2009.

21. Reeve R. 'Wildlife trade, sanctions and compliance: Lessons from the CITES regime' *International Affairs* 2006; 82: 881–897.

22. Ibid.; Bergman, 2009.

23. Humane Society of the United States (HSUS). 'Circus incidents. Attacks, abuse and property damage, revised June, 2004'. www.humanesociety.org, date accessed July 28, 2010.
24. MSNBC. 'Bear trainer distraught after deadly attack' April 4, 2008. www.msnbc.com, date accessed July 17, 2010; Chivvis D. '10 deadly animal attacks' *AOL News* February 25, 2010. www.aolnews.com/world/article/10-deadly-animal-attacks/19374065, date accessed July 24, 2010.
25. *Metro Reporter.* 'Exotic animal-owner mauled to death by his own pet tiger' January 12, 2010. www.metro.co.uk, date accessed June 24, 2010.
26. ABC News. 'Tiger advocate mauled to death by pet' January 12, 2010. www.abc.net.au/news/stories/2010/01/12/2790406.htm, date accessed June 14, 2010.
27. Churney D. 'Wolf attacks girl during show'. Mywebtimes.com April 6, 2009. mywebtimes.com/archives/ottawa/display.php?id=378132&query=wolf%20attacks%20girl%202009 2009, date accessed July 25, 2010.
28. Howard BC. '25 of the worst attacks by exotic pets' *The Daily Green* February 18, 2009. www.thedailygreen.com/living-green/blogs/recycling-design-technology/exotic-pets-attacks-460209, date accessed July 23, 2010.
29. Ibid.
30. Karesh WB, Cook RA, Gilbert M, Newcomb J. 'Implication of wildlife trade on the movement of avian influenza and other infectious diseases' *Journal of Wildlife Diseases* 2007; 43: S55–S59.
31. Gallardo L. 'Exotic pets are popular, but there's a problem' *USA Today* July 27, 2005.
32. Derr, 2002.
33. Boyce G. 'Survivor of chimp attack struggles to live after horrific mauling' *Examiner.com* November 15, 2009. www.examiner.com, date accessed December 12, 2010.
34. Bergman, 2009; Kaur and Aziz, 2006; Adam, 2010; Mee-yoo K. 'Exotic pet trade booming' *Korea Times* September 23, 2009.
35. International Fund for Animal Welfare (IFAW). 'Caught in the web: Wildlife trade on the internet' 2005. www.ifaw.org, date accessed July 20, 2010; BBC News. 'Web trade threat to exotic animals' August 15, 2005. www.news.bbc.co.uk/2/hi/science/nature/4153726.stm, date accessed July 7, 2010.
36. International Animal Rescue. 'Do not encourage wildlife animal trade'. www.internationalanimalrescue.org, date accessed July 8, 2010; Rosenberg M. 'Exotic animals trapped in net of Mexican drug trade' *Reuters* February 5, 2009. www.reuters.com/article/idUSTRE51503V20090206, date accessed June 20, 2010; American Society for the Prevention of Cruelty to Animals (ASPCA). 'Exotic pet trade'. www.aspca.org, date accessed July 8, 2010.
37. Institute of Medicine. *Microbial Threats to Health: Emergence, Detection, and Response* (Washington: National Academies Press) 2003.
38. Newman A. 'Pet chimpanzee attacks woman in Connecticut' *New York Times* February 17, 2009.
39. Schapiro (a). 'All it takes is $45,000 and a phone call to get a pet chimp' *New York Daily News* February 22, 2009.
40. Schapiro (b). 'Mom of crazed chimpanzee, Travis, also shot dead during rage in 2001' *New York Daily News* February 21, 2009.

41. van Hoek CS, ten Cate C. 'Abnormal behavior in caged birds kept as pets' *Journal of Applied Animal Welfare Science* 1998; 1: 51–64.
42. Ibid.; Soulsbury CD, Iossa G, Kennell S, Harris S. 'The welfare and suitability of primates kept as pets' *Journal of Applied Animal Welfare Science* 2009; 12: 1–20; Hobgood J. 'Put a stop to wild animals as pets' *St Petersburg Times (Florida)* July 10, 2009.
43. Thomas S. 'Newton custodian finds python in school locker' *Boston.com* July 7, 2010. www.boston.com, date accessed June 6, 2010.
44. Netter S. 'The allure of exotic pets brings risks to owners, animals' ABC News October 31, 2008. www.abcnews.go.com/US/story?id=6150276 &page=2, date accessed July 10, 2010; Morales T. 'The bad and ugly of exotic pets' CBC News June 25, 2003. www.cbsnews.com, date accessed December 10, 2010.
45. Fox News. 'Miami Zoo takes in unwanted exotic pets on "amnesty" day' February 24, 2008. www.foxnews.com, date accessed July 11, 2010; Netter, 2008; Born Free USA. 'A life sentence: The sad and dangerous realities of exotic animals in private hands' www.bornfreeusa.org, date accessed July 11, 2010; ASPCA. 'Exotic pet trade'.
46. Hartigan M. 'Company to send elephants to home with room to roam' *USA Today* December 7, 2005; Farinato, 2004.
47. Humane Society of the United States (HSUS). 'Circuses'. www.hsus.org, date accessed July 8, 2010; Humane Society of the United States (HSUS). 'Elephant trade factsheet'. www.hsus.org, date accessed March 15, 2010.
48. HSUS, 'Elephant trade factsheet'.
49. Business Wire. 'Christmas comes early for five elephants in South Africa' December 23, 1999. www.allbusiness.com/legal/evidence-testimony/ 6730736-1.html, date accessed August 13, 2010.
50. Boyle C. 'PETA video shows Ringling bros. circus handlers beating elephants' *NY Daily News* July 22, 2009. www. articles.nydailynews.com, date accessed December 13, 2010; Humane Society of the United States (HSUS). 'Ringling Brothers will stand trial for elephant abuse' August 23, 2007. www.hsus.org/wildlife/wildlife_news/ringling_brothers_ trial.html, date accessed July 5, 2010; Humane Society of the United States (HSUS). 'Ringling brothers stand trial in Federal Court for elephant mistreatment' February 2, 2009. www.hsus.org/press_and_publications/press_ releases/ringling_brothers_stands_trial_for_elephant_abuse_020209.html, date accessed July 28, 2010.
51. US Department of Agriculture Inspection. 'Animal and plant inspection service. Inspection Reports. Hendrick Bros. Circus' USDA License #56-C-0121. www.acissearch.aphis.usda.gov/LPASearch/faces/LPASearch.jspx, date accessed December 12, 2010; US Department of Agriculture Inspection. 'Animal and plant inspection service. Inspection Report. Ringling Brothers and Barnum & Bailey Circus' USDA License #52-C-137. www. acissearch.aphis.usda.gov/LPASearch/faces/LPASearch.jspx, date accessed December 12, 2010; US Department of Agriculture Inspection. 'Animal and plant inspection service. Inspection Report. Liebel Family Circus' USDA License #58-C-0288. www.acissearch.aphis.usda.gov/LPASearch/ faces/LPASearch.jspx, date accessed December 12, 2010; US Department of Agriculture Inspection. 'Animal and plant inspection service' APHIS:

AC:WRDeHaven:rf:734-980:5-1-99:c/ac/ringling elephant letter.lwp. www.
acissearch.aphis.usda.gov/LPASearch/faces/LPASearch.jspx, date accessed
December 12, 2010; US Department of Agriculture Inspection. 'Animal and
plant inspection service. Inspection Report. Ringling Bros. and Barnum &
Bailey Circus' USDA License #58-C-106. www.acissearch.aphis.usda.
gov/LPASearch/faces/LPASearch.jspx, date accessed December 12, 2010;
US Department of Agriculture Inspection. 'Animal and plant inspection
service. Inspection Report. Circus Gatti' USDA License #93-C-0792. www.
acissearch.aphis.usda.gov/LPASearch/faces/LPASearch.jspx, date accessed
December 12, 2010; US Department of Agriculture Inspection. 'Animal
and plant inspection service. Inspection Report. Bentley Brothers Circus' USDA License #58-C-0338. www.acissearch.aphis.usda.gov/LPASearch/
faces/LPASearch.jspx, date accessed December 12, 2010; US Department of Agriculture Inspection. 'Animal and plant inspection service.
Inspection Report. Circus Hollywood' USDA License #58-C-0391. www.
acissearch.aphis.usda.gov/LPASearch/faces/LPASearch.jspx, date accessed
December 12, 2010; US Department of Agriculture Inspection. 'Animal and plant inspection service. Inspection Report. Tarzan Zerbini Circus' USDA License #43-C-0012. www.acissearch.aphis.usda.gov/LPASearch/
faces/LPASearch.jspx, date accessed December 12, 2010; 'State of California
Inspection Report. Ringling Bros. and Barnum and Bailey Circus' #A99-
015840. 8-23-99.
52. HSUS, 'Circuses'.
53. Puxley C. 'Ontario readies crackdown on roadside zoos' *The Globe and Mail*
(Canada) March 12, 2008; Dixon J. 'House panel told of abuses by circuses, zoos' *Times Daily* July 8, 1992; Free Lance-Star. 'Roadside zoo animals
starving'. *Free Lance-Star* January 11, 1997.
54. Singer P. 'Free Tilly—And all circus animals' *The Cap Times* March 9, 2010.
55. Hoover W. 'Slain elephant left tenuous legacy in animal rights' *The Honolulu
Advertiser* August 20, 2004. www.the.honoluluadvertiser.com/article/2004/
Aug/20/ln/ln19a.html, date accessed August 13, 2010.
56. MSNBC. 'Why did tiger attack Roy Horn? Case finally closed with no
answers' July 5, 2005. www.today.msnbc.msn.com/id/8391183/ns/today-
entertainment/, date accessed July 11, 2010; CNN. 'Roy of Siegfried and
Roy critical after mauling' *CNN* October 4, 2003.
57. Schapiro (b), 2009.
58. Soulsbury et al., 2009; ASPCA, 'Exotic pet trade'.
59. Harmon K. 'Why would a chimpanzee attack a human?' *Scientific American*
February 19, 2009.
60. Institute of Medicine, 2003.
61. Ibid.; Woolhouse M, Gaunt E. 'Ecological origins of novel human
pathogens' *Critical Reviews in Microbiology* 2007; 33: 231–242.
62. Woolhouse and Gaunt, 2007; Karesh WB, Cook RA, Bennett EL, Newcomb
J. 'Wildlife trade and global disease emergence' *Emerging Infectious Diseases*
2005; 11: 1000–1002.
63. Jones KE, Patel NG, Levy MA et al. 'Global trends in emerging infectious
diseases' *Nature* 2008; 451: 990–994.
64. US Census Bureau. 'U.S. and world population clock'. www.census.gov/
main/www/popclock.html, date accessed November 4, 2011; US Census

Bureau. 'International Database. World population 1950–2050'. www.census.gov/ipc/www/idb/worldpopgraph.php , date accessed July 15, 2010.

65. Vora N. 'Impact of anthropogenic environmental alterations on vector-borne diseases' *Medscape Journal of Medicine* 2008; 10: 238.

66. Aguirre AA, Tabor GM. 'Global factors driving emerging infectious diseases. Impact on wildlife populations' *Annals of the New York Academy of Sciences* 2008; 1149: 1–3; Biotech Business Week. 'Buruli ulcer: Reservoir of deforming tropical disease sought' *Biotech Business Week* November 16, 2009; Molyneux DH. 'Patterns of change in vector-born diseases' *Annals of Tropical Medicine & Parasitology* 1997; 91: 827–839; Patz JA, Daszak P, Tabor GM et al. 'Unhealthy landscapes: Policy recommendations on land use change and infectious disease emergence' *Environmental Health Perspectives* 2004; 112: 1092–1098.

67. Patz et al., 2004; Vora, 2008.

68. Molyneux, 1997.

69. Vora, 2008; Githeko AK, Lindsay SW, Confalonieri UE, Patz JA. 'Climate change and vector-borne diseases: A regional analysis' *Bulletin of the World Health Organization* 2000; 78: 1136–1147.

70. Patz et al., 2004.

71. Torres-Velez F, Brown C. 'Emerging infections in animals—Potential new zoonoses?' *Clinics in Laboratory Medicine* 2004; 24: 825–838.

72. Greger M. 'The human/animal interface: Emergence and resurgence of zoonotic infectious diseases' *Critical Reviews in Microbiology* 2007; 33: 243–299; Torres-Velez and Brown, 2004.

73. Woolhouse MEJ. 'Population biology of emerging and re-emerging pathogens' *Trends in Microbiology* 2002; 10 (10 Suppl.): S3–S7.

74. Lancet Editorial. 'Avian influenza: the threat looms' *Lancet* 2004; 363: 257.

75. Calvert S, Kohn D. 'Out of Africa: A baffling variety of diseases' *The Baltimore Sun* May 15, 2005.

76. Wolfe ND, Switzer WM, Carr JK et al. (a). 'Naturally acquired simian retrovirus infections in central African hunters' *Lancet* 2004; 363: 932–937.

77. Merson MH. 'The HIV-AIDS pandemic at 25—The global response' *New England Journal of Medicine* 2006; 354: 2414–2417.

78. Bell D, Roberton S, Hunter PR. 'Animal origins of SARS coronavirus: Possible links with the international trade in small carnivores' *Philosophical Transactions of the Royal Society of London B Biological Sciences* 2004; 359: 1107–1114.

79. Barry E. 'A taste of baboon and monkey meat, and maybe of prison, too' *New York Times* November 17, 2004; Doward J. 'Virus fear over smuggled bushmeat: Diseases that pose a threat to humans, such as Ebola, may be entering UK through the illegal food trade' *Observer* May 8, 2005.

80. Fletcher M. 'Extinction looms in a paradise lost to guns, greed, and hunger; the endangered black rhino and other wildlife are being massacred as gangs take advantage of lawlessness and poverty, writes Martin Fletcher' *The Times* December 18, 2008; Mainka SA, Mills JA. 'Wildlife and traditional Chinese medicine—Supply and demand for wildlife species' *Journal of Zoo and Wildlife Medicine* 1995; 26: 193–200; Gratwicke B, Mills J, Dutton A et al. 'Attitudes toward consumption and conservation of tigers in China' *PLoS One* 2008; 3: e2544; Alacs E, Georges A. 'Wildlife across our borders:

A review of the illegal trade in Australia' *Australian Journal of Forensics Sciences* 2008; 40: 147–160.

81. Humane Society of the United States (HSUS). 'The unbearable trade in bear parts and bile'. www.hsus.org, date accessed March 15, 2010; Dinerstein E, Loucks C, Wikramanayake E. et al. 'The fate of wild tigers' *Bioscience* 2007; 57: 508–514; Telecky T. 'Illegal tiger trade must end' *Humane Society of the United States* January 16, 2008. www.hsus.org, date accessed July 17, 2010; Gratwicke et al., 2008; Sullivan, 2003; Bauer AM. 'Geckos in traditional medicine: Forensic implications' *Applied Herpetology* 2009; 6: 81–96.

82. Karesh WB, Noble E. 'The bushmeat trade: Increased opportunities for transmission of zoonotic disease' *Mount Sinai Journal of Medicine* 2009; 76: 429–434; Humane Society of the United States (HSUS). 'Bushmeat'. www. hsus.org, date accessed March 15, 2010; Bushmeat Crisis Task Force (BCTF). 'Economics of bushmeat'. www.bushmeat.org, date accessed March 15, 2010.

83. Suarez E, Morales M, Cueva R et al. 'Oil industry, wild meat trade and roads: Indirect effects of oil extraction activities in a protected area in north-eastern Ecuador' *Animal Conservation* 2009; 12: 364–373; Crookes DJ, Milner-Gulland EJ. 'Wildlife and economic policies affecting the bushmeat trade: A framework for analysis' *South African Journal of Wildlife Research* 2006; 36: 159–165; Bushmeat Crisis Task Force (BCTF). 'A wildlife crisis in west and central Africa and around the world'. www.bushmeat.org, date accessed March 15, 2010; Zhang L, Hua N, Sun S. 'Wildlife trade, consumption and conservation awareness in southwest China' *Biodiversity and Conservation* 2008; 17: 1493–1516; Sullivan, 2003.

84. Thibault M, Blaney S. 'The oil industry as an underlying factor in the bushmeat crisis in central Africa' *Conversation Biology* 2003; 17: 1807–1813; Poulson JR, Clark CJ, Mavah G, Elkan PW. 'Bushmeat supply and consumption in a tropical logging concession in northern Congo' *Conservation Biology* 2009; 23: 1597–1608; Wilkie DS, Bennett EL, Eves HE, Hutchins M, Wolf C. 'Roots of the bushmeat crisis: Eating the world's wildlife into extinction' *AZA Communiqué* November 6–7, 2002; Karesh and Noble, 2009.

85. BCTF. 'A wildlife crisis in west...'

86. Science Daily. 'DNA barcodes: Creative new uses span health, fraud, smuggling, history, more' *Science Daily* November 12, 2009. www.sciencedaily. com, date accessed December 3, 2009.

87. Karesh and Noble, 2009; Wilkie et al., 2002; Nielsen J. 'Confronting central Africa's poaching crisis. International efforts target abuses of "bushmeat trade"' *National Public Radio* January 12, 2003. www.npr.org/templates/story/story.php?storyId=912962, date accessed June 12, 2010.

88. Bowen-Jones E, Brown D, Robinson EJZ. 'Economic commodity or environmental crisis? An interdisciplinary approach to analyzing the bushmeat trade in central and West Africa' *Area* 2003; 35: 390–402.

89. Wolfe ND. 'Preventing the next pandemic' *Scientific American* 2009; 300: 76–81.

90. Humane Society of the United States (HSUS). 'Wildlife trade'. www.hsus. org, date accessed March 15, 2010.

91. Ecorazzi. 'Skin trade documentary offers compelling case for passing on fur altogether' *Ecorazzi* June 15, 2010. www.ecorazzi.com, date accessed July 8, 2010; Moore P. 'No crocodile tears for tanking skins trade' *American Chronicle* December 4, 2009.

92. Rawstorne T. 'Pythons skinned and left to die. The shocking reality behind fashion's new obsession' *MailOnline* September 20, 2007. www.dailymail.co. uk/femail/article-482849/Pythons-skinned-left-die-The-shocking-reality-fashions-new-obsession.html, date accessed July 8, 2010.

93. Jacmenovic M. 'Southern China is not the only source for volatile live markets' *Humane Society of the United States*. www.hsus.org, date accessed March 15, 2010.

94. Gao F, Bailes E, Robertson DL et al. 'Origin of HIV-1 in the chimpanzee *Pan troglodytes troglodytes*' *Nature* 1999; 397: 436–441

95. Ibid.; Sharp PM, Bailes E, Chaudhuri RR, Rodenburg CM, Santiago MO, Hahn BH. 'The origins of acquired immune deficiency syndrome viruses: where and when?' *Philosophical Transactions of the Royal Society London B Biological Sciences* 2001; 356: 867–876; Van Heuverswyn F, Peeters M. 'The origins of HIV and implications for the global epidemic' *Current Infectious Disease Reports* 2007; 9: 338–346.

96. Peeters M, Courgnaud V, Abela B et al. 'Risk to human health from a plethora of simian immunodeficiency viruses in primate bushmeat' *Emerging Infectious Diseases* 2002; 8: 451–457.

97. Wolfe ND, Posser AT, Carr JK et al. (b). 'Exposure to nonhuman primates in rural Cameroon' *Emerging Infectious Diseases* 2009; 10: 2094–2099.

98. Drexler M. *Emerging Epidemics* (New York: Pengiun Books) 2002, p. 8.

99. Woolhouse and Gaunt, 2007.

100. Crawford D. *The Invisible Enemy: A Natural History of Viruses* (New York: Oxford University Press) 2003, p. 2.

101. Wolfe ND, Heneine W, Carr JK et al. (a). 'Emergence of unique primate T-lymphotropic viruses among central African bushmeat hunters' *Proceedings of the National Academy of Sciences* 2005; 102: 7994–7999.

102. Ibid.

103. Wolfe et al. (a), 2004.

104. Switzer WM, Bhullar V, Shanmugam V et al. 'Frequent simian foamy virus infection in persons occupationally exposed to nonhuman primates' *Journal of Virology* 2004; 78: 2780–2789; Lerche NW, Switzer WM, Yee JL et al. 'Evidence of infection with simian type D retrovirus in persons occupationally exposed to nonhuman primates' *Journal of Virology* 2001; 75: 1783–1789; Sandstrom PA, Phan KO, Switzer WM et al. 'Simian foamy virus infection among zoo keepers' *Lancet* 2000; 355: 551–552.

105. Wolfe ND, Eitel MN, Gockowski J et al. 'Deforestation, hunting and the ecology of microbial emergence' *Global Change and Human Health* 2000; 1: 10–25; Wolfe, 2009.

106. Goldberg TL, Chapman CA, Cameron K et al. 'Serologic evidence for novel poxvirus in endangered red colubus monkeys in western Uganda' *Emerging Infectious Diseases* 2008; 14: 801–803.

107. Rouquet P, Froment JM, Bermejo M et al. 'Wild animal mortality monitoring and human Ebola outbreaks, Gabon and Republic of Congo, 2001–2003' *Emerging Infectious Diseases* 2005; 11: 283–290.

108. Groseth A, Feldmann H, Strong JE. 'The ecology of Ebola virus' *Trends in Microbiology* 2007; 15: 408–416; Leroy EM, Kumulungui B, Pourrut X et al. 'Fruit bats as reservoirs of Ebola virus' *Nature* 2005; 438: 575–576.
109. Leroy et al., 2005; Morvan JM, Deubel V, Gounon P et al. 'Identification of Ebola virus sequences present as RNA or DNA in organs of terrestrial small mammals of the Central African Republic' *Microbes and Infection* 1999; 1: 1193–1201; Rouquet et al., 2005.
110. Wolfe et al. (a), 2004; Wolfe et al. (a), 2005.
111. Le Breton M, Yang O, Tamoufe U et al. 'Exposure to wild primates among HIV-infected persons' *Emerging Infectious Diseases* 2007; 13: 1579–1582; Goldberg et al., 2008.
112. Le Breton et al., 2007.
113. Peeters et al., 2002.
114. Murray K, Selleck P, Hooper P et al. 'A morbillivirus that caused fatal disease in horses and humans' *Science* 1995; 268: 94–97; Centers for Disease Control and Prevention (CDC). 'Hendra Virus disease and Nipah Virus encephalitis'. www.cdc.gov, date accessed July 21, 2010.
115. Selvey L, Taylor R, Arklay A, Gerrard J. 'Screening of bat carriers for antibodies to equine morbillivirus' *Communicable Diseases Intelligence* 1996; 20: 477–478.
116. Greger, 2007.
117. Gould AR. 'Comparison of the deduced matrix and fusion protein sequences of equine morbillivirus with cognate genus of the Paramyxoviridae' *Virus Research* 1996; 43: 17–31; Greger, 2007.
118. Greger, 2007; Murphy FA. 'Emerging zoonoses: The challenge for public health and biodefense' *Preventive Veterinary Medicine* 2008; 86: 216–223; Centers for Disease Control and Prevention (CDC). 'Severe acute respiratory syndrome'. www.cdc.gov, date accessed July 19, 2010.
119. Hepeng J. 'China culls civets in bid to stop SARS'. *SciDev.net* January 6, 2004. www.scidev.net/en/news/china-culls-civets-in-bid-to-stop-sars.html, date accessed August 10, 2010; Li W, Shi Z, Yu M et al. 'Bats are natural reservoirs of SARS-like coronaviruses' *Science* 2005; 310: 676–679; Fevre EM, de C. Bronsvoort BM, Hamilton K, Cleaveland S. 'Animal movements and the spread of infectious diseases' *Trends in Microbiology* 2006; 14: 125–131; Lau SK, Woo PC, Li KS et al. 'Severe acute respiratory syndrome coronavirus-like virus in Chinese horseshoe bats' *Proceedings of the National Academy of Sciences* 2005; 102: 14040–14045.
120. Dove J. 'Animals die in live-animal markets' *Earth Island Journal* Winter/Spring 1998–99; 1; Jacmenovic, 2010.
121. Jacmenovic, 2010; Benatar D. 'The chickens come home to roost' *American Journal of Public Health* 2007; 97: 1545–1546.
122. Phillips G. 'Activists point to widespread animal abuse' *Taipei Times* March 27, 2005.
123. Dove, 1999.
124. Warren S, Auster L. 'China's animal torture' *FrontPage Magazine* April 26, 2006.
125. Woo PC, Lau SK, Yeun K-Y. 'Infectious diseases emerging from Chinese wet-markets: Zoonotic origin of severe respiratory viral infections' *Current Opinion in Infectious Diseases* 2006; 19: 401–407.

126. Fevre et al., 2006; Li et al., 2005; Lau et al., 2005.
127. Netting JF. 'Are bats the source of SARS?' *Discover* December 1, 2005.
128. Ibid.
129. Benatar, 2007.
130. Centers for Disease Control and Prevention (CDC). 'Known cases and out-breaks of Ebola hemmorhagic fever, in chronological order'. www.cdc.gov, date accessed June 20, 2010.
131. Wolfe ND, Daszak P, Kilpatrick AM, Burke DS. (b). 'Bushmeat hunting, deforestation, and prediction of zoonoses emergence' *Emerging Infectious Diseases* 2005; 11: 1822–1827.
132. Greger, 2007; Li et al., 2005.
133. Wolfe et al. (b), 2005.
134. World Health Organization (WHO). 'Marburg hemorrhagic fever fact sheet'. www.who.int/csr/disease/marburg/factsheet/en/, date accessed July 18, 2010.
135. Chomel BB, Belotto A, Meslin F-X. 'Wildlife, exotic pets, and emerging zoonoses' *Emerging Infectious Diseases* 2007; 13: 6–11.
136. Marchione M. 'Tighter rules sought on exotic pets after trade, risks increase' *Milwaukee Sentinel Journal* August 6, 2003.
137. Humane Society of the United States (HSUS). 'Live cargo'. www.hsus,org, date accessed March 15, 2010; Wyler LS, Sheikh PA. 'International ille-gal trade in wildlife: Threats and U.S. Policy' *CRS Report for Congress*, March 3, 2008. fpc.state.gov/documents/organization/102621.pdf, date accessed June 29, 2010.
138. Wyler and Sheikh, 2008; HSUS, 'Live cargo'; Bergman, 2009; Alacs and Georges, 2008.
139. Adam, 2010; Wasser SK, Clark WJ, Drori O et al. 'Combating the illegal trade in African elephant ivory with DNA forensics' *Conservation Biology* 2008; 22: 1065–1071; Schmidt CW. 'Crimes. Earth's expense' *Environmental Health Perspectives* 2004; 1121: A97–A103; Wyler and Sheikh, 2008.
140. Wyler and Sheikh, 2008.
141. Ibid.; Schmidt, 2004; Baker, 2008; Derr, 2002.
142. Wyler and Sheikh, 2008; Alacs and Georges, 2008; Wasser SK, Mailand C, Booth R et al. 'Using DNA to track the origin of the largest ivory seizure since the 1989 ban' *Proceedings of the National Academy of Sciences* 2007; 104: 4228–4223.
143. Bergman, 2009.
144. Rosen GE, Smith KF. 'Summarizing the evidence on the international trade in illegal wildlife' *Ecohealth* June 4, 2010 (Epub ahead of print); Wyler and Sheikh, 2008; Sullivan, 2003; TRAFFIC. 'The State of wildlife trade in China. Information on the trade in wild animals and plants in China, 2008'. www.traffic.org, date accessed November 24, 2010; Adam, 2010.
145. Bergman, 2009; Derr, 2002; HSUS, 'Wildlife trade'.
146. HSUS, 'Live cargo'.
147. Humane Society of the United States (HSUS). 'The trade in live reptiles: Exports from the United States'. www.hsus,org, date accessed April 20, 2010; Bilger, 2009.
148. Humane Society of the United States (HSUS). 'The trade in live reptiles: Imports to the United States'. www.hsus,org, date accessed April 20, 2010.

149. HSUS, 'Wildlife trade'.
150. Bilger, 2009; Derr, 2002.
151. HSUS, 'The trade in live reptiles: Imports to the United States'; HSUS, 'Live cargo'.
152. Netter, 2008; Kaur and Aziz, 2006.
153. Bergman, 2009.
154. Rosenberg, 2009.
155. Bergman, 2009.
156. Rosenberg, 2009.
157. Murphy, 2008.
158. Greger, 2007.
159. Karesh WB, Cook RA. 'One world—One health' *Journal of the Royal College of Physicians* 2009; 9: 259–260.
160. World Health Organization (WHO). 'Smallpox. Historical significance'. www.who.int/mediacentre/factsheets/smallpox/en/, date accessed August 10, 2010.
161. Woodyard C. 'Virus raises issue: Pet or threat?' *USA Today* June 10, 2003.
162. Stephenson J. 'Monkeypox outbreak a reminder of emerging infections vulnerabilities' *Journal of the American Medical Association* 2003; 290: 23–24; Marano N, Arguin PM, Pappaioanou M. 'Impact of globalization and animal trade on infectious disease ecology' *Emerging Infectious Diseases* 2007; 13: 1807–1808.
163. Marchione, 2003.
164. Torres-Valez and Brown, 2004.
165. Orent W. 'Will monkeypox be the next smallpox?' *Los Angeles Times* September 26, 2010.
166. Harris JR, Bergmire-Sweat D, Schlegal JH et al. 'Multistate outbreaks associated with small turtle exposure, 2007–2008' *Pediatrics* 2009; 124: 1388–1394; Centers for Disease Control and Prevention (CDC). 'Reptile-associated salmonellosis—Selected states, 1998–2002' *Morbidity and Mortality Weekly Report* 2003; 52: 1206–1209; Centers for Disease Control and Prevention (CDC). 'Reptiles, amphibians and salmonella'. www.cdc.gov, date accessed July 23, 2010.
167. Mermin J, Hutwagner L, Vugia D et al. 'Reptiles, amphibians, and human Salmonella infection: A population-based, case-control study' *Clinical Infectious Diseases* 2004; 38: S253–S261; Chomel BB, Belotto A, Meslin F-X. 'Wildlife, exotic pets, and emerging zoonoses' *Emerging Infectious Diseases* 2007; 13: 6–11.
168. Harris JR, Neil KP, Behravesh CB, Sotir MJ, Angulo FJ. 'Recent multistate outbreaks of human salmonella infections acquired from turtles: A continuing public health challenge' *Clinical Infectious Diseases* 2010; 50: 554–559.
169. CDC, 'Reptiles, amphibians and salmonella'; Harris et al., 2009.
170. Harris et al., 2010; CDC, 'Reptiles, amphibians and salmonella'.
171. Geu L, Loschner U. 'Salmonella enterica reptiles of German and Austrian origin' *Veterinary Microbiology* 2002; 84: 79–91.
172. Nakadai A, Kuroki T, Kato Y et al. 'Prevalence of Salmonella spp. in pet reptiles in Japan' *Journal of Veterinary Medical Science* 2005; 67: 97–101.
173. CDC, 'Reptiles, amphibians and salmonella'.

174. Friedman CR, Torigian C, Shilam PJ et al. 'An outbreak of salmonellosis among children attending a reptile exhibit at a zoo' *Journal of Pediatrics* 1998; 132: 802–807.
175. Centers for Disease Control and Prevention (CDC). 'Multistate outbreak of human *Salmonella* Typhimurium infections associated with pet turtle exposure—United States, 2008' *Morbidity and Mortality Weekly Report* 2010; 59: 191–196; CDC, 2003.
176. Bertrand S, Rimhanen-Finne R, Weill FX et al. 'Salmonella infections associated with reptiles: The current situation in Europe' *Eurosurveillance* 2008; 13: 1–6; Tauxe RV, Rigau-Perez JG, Wells JG, Blake PA. 'Turtle-associated salmonellosis in Puerto-Rico. Hazards of the global turtle trade' *Journal of the American Medical Association* 1985; 254: 237–239.
177. Riley PY, Chomel BB. 'Hedgehog zoonoses' *Emerging Infectious Diseases* 2005; 11: 1–5; Ostrowski SR, Leslie MJ, Parrott T, Albet S, Piercy PE. 'B-virus from pet macaque monkeys: An emerging threat in the United States?' *Emerging Infectious Diseases* 1998; 4: 117–121; Mcleod L. 'Health Canada advisory about pet hamsters' *About.com* November 16, 2004. www.exoticpets.about.com/od/healthandsafetyissues/ a/hamsterstularem.htm, date accessed July 20, 2010; Dorrestein GM, Wiegman LJ. 'Inventory of the shedding of *Chlamydia psittaci* by parakeets in the Utrecht area using ELISA' *Tijdschr Diergenesskd* 1989; 15: 1227–1236; Schmiedeknecht G, Eikmann M, Kohler K et al. 'Fatal cowpox virus infection in captive branded mongooses (mungos mungo)' *Veterinary Pathology Online* 2010; 47: 547.
178. Inoue K, Maruyama S, Kabeya H et al. 'Exotic small mammals as potential reservoirs of zoonotic *Bartonella* spp' *Emerging Infectious Diseases* 2009; 15: 526–532.
179. Moroney JF, Guevara R, Iverson C et al. 'Detection of chlamydiosis in shipment of pet birds, leading to recognition of an outbreak of clinically mild psittacosis in humans' *Clinical Infectious Diseases* 1998; 26: 1425–1429.
180. Michalak K, Austin C, Diesel S, Bacon JM, Zimmerman P, Maslow JN. '*Mycobacterium tuberculosis* infection as a zoonotic disease: Transmission between humans and elephants' *Emerging Infectious Disease* 1998; 4: 283–287.
181. Viegas J. 'Parasites spreading between animals, zookeepers. The finding suggests cross-species infections may be common in zoos world-wide' *Discovery News* January 25, 2010.
182. Stetter MD, Mikota SK, Gutter AF et al. 'Epizootic of *Mycobacterium bovis* in a zoological park' *Journal of the American Veterinary Medical Association* 1995; 201: 1618–1621.
183. Viegas, 2010.
184. Bender JB, Shulman SA. 'Reports of zoonotic disease outbreaks associated with animal exhibits and availability of recommendations for preventing zoonotic disease transmission from animals to people in such settings' *Journal of the American Veterinary Medical Association* 2004; 224: 1105–1109.
185. Pavlin BI, Schloegel LM, Daszak P. 'Risks of importing zoonotic diseases through wildlife trade, United States' *Emerging Infectious Diseases* 2009; 15: 1721–1726.
186. Marchione, 2003.

187. Vie JC, Hilton-Taylor C, Stuart ST (eds) *Wildlife in a changing world—An analysis of the 2008 IUCN Red List of threatened species* (Gland, Switzerland: IUCN) 2009.
188. Lee RL, Gorog AJ, Dwiyahreni A et al. 'Wildlife trade and implications for law enforcement in Indonesia: A case study from North Sulawesi' *Biological Conservation* 2005; 123: 477–488.
189. MacFarqhar N. 'U.N. sets goals to reduce the extinction rate' *New York Times* October 29, 2010.
190. The World Conservation Union. 'IUCN Red List of Threatened Species. Summary statistics for globally threatened species. Table 1: Numbers of threatened species by major groups of organisms (1996–2010), 2010'. www.iucnredlist.org/about/summary-statistics, date accessed October 20, 2010.
191. Marris E. 'Bagged and boxed: It's a frog's life' *Nature* 2008; 452: 394–395.
192. WildAid. 'The illegal wildlife trade'. www.wildaid.org, date accessed July 1, 2010 ; Telecky, 2008.
193. Bergman, 2009.
194. Jha A. 'Almost half of all primates face "imminent extinction" *Guardian* February 18, 2010.
195. Randerson J. 'Nearly half of all the world's primates at risk of extinction' *Guardian* August 5, 2008.
196. HSUS, 'Elephant trade factsheet'; Wasser SK, Clark WJ, Drori O et al. 'Combating the illegal trade in African elephant ivory with DNA forensics' *Conservation Biology* 2008; 22: 1065–1071; Blake S, Strindberg S, Boudjan P et al. 'Forest elephant crisis in the Congo Basin' *PLoS Biology* 2007; 5: 0945–0953.
197. Leakey R, Lewin R. *The Sixth Extinction: Patterns of Life and the Future of Humankind* (New York: Doubleday, Anchor) 1996.
198. Anonymous. 'The sixth extinction'. www.well.com/~davidu/sixthextinction.html, date accessed January 14, 2011.
199. Warrick J. 'Mass extinction underway, majority of biologists say' *Washington Post* April 21, 1998.
200. Kock R, Kebkiba B, Heinonen R, Bedane B. 'Wildlife and pastoral society-shifting paradigms in disease control' *Annals of the New York Academy of Sciences* 2002; 969: 24–33.
201. Vieira IC, Toledo PM, Silva JM, Higuchi H. 'Deforestation and threats to biodiversity of Amazonia' *Brazilian Journal of Biology* 2008; 68 (Suppl. 4): 949–956; Lindenmayer DB. 'Forest wildlife management and conservation' *Annals of the New York Academy of Science* 2009; 1162: 283–310.
202. Lindenmayer, 2009; Yiming L, Wilcove DS. 'Threats to vertebrate species in China and the United States' *Bioscience* 2005; 55: 147–153; Jha S, Bawa KS. 'Population growth, human development, and deforestation in biodiversity hotspots' *Conservation Biology* 2006; 20: 906–912; Collins JP, Storfer A. 'Global amphibian declines: Sorting the hypotheses' *Diversity and Distributions* 2003; 9: 89–98; Lindenmayer, 2009; Young BE, Lips KR, Reaser JK et al. 'Population declines and priorities for amphibian conservation in Latin America' *Conservation Biology* 2001; 15: 1213–1223.
203. Yiming and Wilcove, 2005; Adam, 2010; Missios PC. 'Wildlife trade and endangered species protection' *Australian Journal of Agricultural and Resource Economics* 2004; 48: 613–627; Bowen-Jones et al., 2003; Thibault M,

Blaney S. 'The oil industry as an underlying factor in the bushmeat crisis in central Africa' *Conversation Biology* 2003; 17: 1807–1813.

204. Bushmeat Crisis Task Force (BCTF). 'Social ecology fact sheet'. www. bushmeat.org, date accessed March 15, 2010.

205. Derr, 2002.

206. Hood M. 'UN wildlife body rejects bid to reopen ivory trade' *Wild Singapore News* March 22, 2010.

207. Randerson, 2008.

208. Rivalin P, Delmas V, Angulo E et al. 'Can bans stimulate wildlife trade?' *Nature* 2007; 447: 529–530.

209. Angulo E, Deves AL, Saint Jalmes M, Courchamp F. 'Fatal attraction: Rare species in the spotlight' *Proceedings of the Royal Society B Biological Sciences* 2009; 276: 1331–1337; Adam, 2010.

210. Adam, 2010.

211. Sullivan, 2003; Derr, 2002

212. Bergman, 2009.

213. HSUS, 'The trade in live reptiles: Imports to the United States'.

214. Nielsen, 2003.

215. Bennett EL, Robinson JG. *Hunting of Wildlife in Tropical Forests. Biodiversity Series-Impact Studies, Paper 76* (Washington, DC: World Bank) 2000.

216. Lee et al., 2005; Wilkie et al., 2002; Barrett MA, Ratsimbazafy J. 'Luxury bushmeat trade threatens lemur conservation' *Nature* 2009; 461: 470.

217. Bennett and Robinson, 2000; Wilkie et al., 2002; HSUS, 'The trade in live reptiles: Exports from the United States'.

218. Bennett and Robinson, 2000.

219. Redford KH. 'The empty forest' *Bioscience* 1992; 42: 412–422.

220. Wyler and Sheikh, 2008; Carpenter et al., 2004; Rosen and Smith, 2010.

221. Russello MA, Avery ML, Wright TF. 'Genetic evidence links invasive monk parakeet populations in the United States to the international pet trade' *BMC Evolutionary Biology* 2008; 8: 217–227; Fox News, 2008; Netter, 2008; Hobgood J. 'Put a stop to wild animals as pets' *St Petersburg Times* (Florida) July 10, 2009; Bilger, 2009.

222. Russello et al., 2008; Aldhous P. 'Exotic pets pose risks to native species' *New Scientist* August 1, 2007. www.newscientist.com, date accessed June 21, 2010; Hobgood, 2009.

223. Hobgood, 2009; Bilger, 2009.

224. HSUS, 'The trade in live reptiles: Imports to the United States'; Karesh WB, Cook RA, Gilbert M, Newcomb J. 'Implication of wildlife trade on the movement of avian influenza and other infectious diseases' *Journal of Wildlife Diseases* 2007; 43: S55–S59.

225. Marris, 2008; Picco AM, Collins JP. 'Amphibian commerce as a likely source of pathogen pollution' *Conservation Biology* 2008; 22: 1582–1589.

226. Schloegel LM, Picco AM, Kilpatrick AM, Davies AJ, Hyatt AD, Daszak P. 'Magnitude of the US trade in amphibians and presence of *Batrachochytrium dendrobatidis* and ranavirus infection in imported North American bullfrogs (*Rana catesbeiana*)' *Biological Conservation* 2009; 142: 1420–1426.

227. Thompson RCA, Kutz SJ, Smith A. 'Parasite zoonoses and wildlife: Emerging issues' *International Journal of Environmental Research and Public Health* 2009; 6: 678–693.
228. Adam, 2010.
229. Bergman, 2009; Conservation International. 'Biodiversity hotspots'. www. biodiversityhotspots.org, date accessed July 3, 2010.
230. Tuhus-Dubrow R. 'Martial law of the jungle. When defending the environment means calling in the military' *The Boston Globe* December 21, 2008.
231. Derr, 2002.
232. Ostfeld RS. 'Biodiversity loss and the rise of zoonotic pathogens' *Clinical Microbiology and Infection* 2009; 15 (Suppl. 1): 40–43.
233. Ibid.
234. Centers for Disease Control and Prevention (CDC). 'Lyme disease—United States, 2003–2005' *Morbidity and Mortality Weekly Report* 2007; 56: 573–576.
235. Ostfeld, 2009.
236. Ostfeld RS, Keesing F. 'Biodiversity and disease risk: The case of Lyme disease' *Conservation Biology* 2000; 14: 722–728.
237. Otsfeld, 2009; Ezenwa VO, Godsey MS, King RJ, Guptill SC. 'Avian diversity and West Nile virus: Testing associations between biodiversity and infectious disease risk' *Proceedings of the Royal Society B Biological Sciences* 2006; 273: 109–117; Mills JN. 'Biodiversity loss and emerging infectious disease: An example from rodent-borne hemorrhagic fevers' *Biodiversity* 2006; 7: 9–17; Chivian E (ed.) *Biodiversity: Its Importance to Human Health. Interim Executive Summary. Center for Health and Global Environment, Harvard Medical School 2002.* chge.med.harvard.edu/publications/documents/Biodiversity_v2_screen.pdf, date accessed July 6, 2010.
238. Chivian, 2002.
239. Gerson H, Cudmore B, Mandrak NE, Coote LD, Farr K, Baillargeon G. 'Monitoring international wildlife trade with coded species data' *Conservation Biology* 2008; 22: 4–7.
240. Weinhold B. 'Infectious disease: The human costs of our environmental errors' *Environmental Health Perspective* 2004; 112: A32–A39.
241. Derr, 2002; Baker, 2008.
242. Reaser JK, Clark Jr EE, Meyers NM. 'All creatures great and minute: A public policy primer for companion animal zoonoses' *Zoonoses and Public Health* 2008; 385–401.
243. Marano et al., 2007.
244. Ibid.
245. HSUS, 'The trade in live reptiles: Imports to the United States'.
246. Bergman, 2009.
247. Messmer TO, Skelton SK, Moroney JF, Daugharty H, Fields BS. 'Application of a nested, multiplex PCR to psittacosis outbreaks' *Journal of Clinical Microbiology* 1997; 35: 2043–2046.
248. Moroney et al., 1998.
249. Ostrowski et al., 1998.
250. Centers for Disease Control and Prevention (CDC). 'Multistate outbreak of human salmonella infections associated with exposure to turtles—United

States, 2007–2008' *Morbidity and Mortality Weekly Report* 2008; 57: 69–72.

251. Shane SM, Gilbert R, Harrington KS. 'Salmonella colonization in commercial pet turtles (*Pseudemys scripta elegans*)' *Epidemiology and Infection* 1990; 105: 307–316; D'Aoust JY, Daley E, Crozier M, Sewell AM. 'Pet turtles: A continuing international threat to public health' *American Journal of Epidemiology* 1990; 132: 233–238; Siebeling RJ, Caruso D, Neuman S. 'Eradication of *Salmonella* and *Arizona* species from turtle hatchlings produced from eggs treated on commercial turtle farms' *Applied and Environmental Microbiology* 1984; 47: 658–662; Izadjoo MJ, Pantoja CO, Siebeling RJ. 'Acquisition of *Salmonella* flora by turtle hatchlings on commercial turtle farms' *Canadian Journal of Microbiology* 1987; 33: 718–724; Tauxe et al., 1985.
252. Shane et al., 1990.
253. Díaz MA, Cooper RK, Cloeckaert A, Siebeling RJ. 'Plasmid mediated high-level gentamicin resistance among enteric bacteria isolate from pet turtles in Louisians' *Applied and Environmental Microbiology* 2006; 72: 306–312; D'Aoust et al., 1990.
254. Brenner D, Lewbart G, Stebbins M et al. 'Health survey of wild and captive bog turtles (*Clemmys muhlenbergiii*) in North Carolina and Virginia' *Journal of Zoo and Wildlife Medicine* 2002; 33: 311–316; Mitchell JC, McAvoy BV. 'Enteric bacteria in natural populations of freshwater turtles in Virginia' *Virginia Journal of Science* 1990; 41: 233–242; Richards JM, Brown JD, Kelly TR et al. 'Absence of detectable *Salmonella* cloacal shedding in free-living reptiles on admission to the wildlife center of Virginia' *Journal of Zoo and Wildlife Medicine* 2004; 35: 562–563; Saelinger CA, Lewbart GA, Christian LS, Lemons CL. 'Prevalence of *Salmonella* spp in cloacal, fecal, and gastrointestinal mucosal samples from wild North American turtles' *Journal of the American Veterinary Medical Association* 2006; 229: 266–268.
255. Saelinger et al., 2006.
256. Ferenczi D. 'Frogs blamed for salmonella outbreak' *Consumer Reports* April 15, 2011.
257. TRAFFIC, 2008.
258. Dinerstein et al., 2007.
259. TRAFFIC, 2008.
260. HSUS, 'The trade in live reptiles: Exports from the United States'.
261. TRAFFIC, 2008; Haitao P, Parham JF, Lau M, Tien-Hsi C. 'Farming endangered turtles to extinction in China' *Conservation Biology* 2007; 21: 5–6.
262. Telecky, 2008; Zhou and Zhang, 2008; TRAFFIC, 2008.
263. TRAFFIC, 2008; Telecky, 2008; Gratwicke et al., 2008.
264. TRAFFIC, 2008; McAllister RRJ, McNeill D, Gordon IJ. 'Legalizing markets and the consequences for poaching of wildlife species: The vicuna as a case study' *Journal of Environmental Management* 2009; 90: 120–130.
265. Begley S, Johnson S, Overdorf J. 'Extinction trade; Endangered animals are the new blood diamonds as militias and warlords use poaching to fund death' *Newsweek* March 10, 2008.
266. Sullivan, 2003.
267. Wyler and Sheikh, 2008; Begley et al., 2008.

268. Marchione M. 'Globetrotting boosts exotic diseases' *Milwaukee Journal Sentinel* June 15, 2003.
269. Reid M. 'Exotic pet trade is "raising risk of global pandemic"' *The Times* October 29, 2007.
270. Pitman T. 'In Congo forest, bushmeat trade threatens Pygmies' *MSNBC* July 3, 2010. www.msnbc.com, date accessed December 12, 2010.
271. Tobiason A. 'Thinking BIG: The Congo Basin Forest partnership' *Bushmeat Crisis Task Force* June 12, 2009. www.bushmeat.org, date accessed December 12, 2010.
272. Namibia Government. 'Introduction to Namibia'. www.namibia government.com/communal_wildlife_concervancies.htm, date accessed December 12, 2010; Naidoo R, Adamowicz WL. 'Economic benefits of biodiversity exceed costs of conservation at an African rainforest reserve' *Proceedings of the National Academy of Sciences* 2005; 102:16712–16716; Kirkby CA, Giudice-Granados R, Day B et al. 'The market triumph of ecotourism: An economic investigation of the private and social benefits of competing land uses in the Peruvian Amazon' *Public Library of Science One* September 29, 2010; 5 (9): e13015.
273. Tobiason, 2009.
274. Wolfe, 2009.

4 Foul Farms: The State of Animal Agriculture

1. Barrette RW, Metwally SA, Rowland JM et al. 'Discovery of swine as a host for the *Reston ebolavirus*' *Science* 2009; 325: 204–206; Normille D. 'Scientists puzzle over Ebola-Reston virus in pigs' *Science* 2009; 323: 451; Cyranoski D. 'Ebola outbreak has experts rooting for answers' *Nature* 2009; 457: 364–365.
2. Barrette et al., 2009.
3. Cyranoski, 2009.
4. Pearson J, Salman MD, BenJabara K et al. 'Global risks of infectious animal diseases' Council for Agricultural Science and technology, Issue Paper No. 28, 2005.
5. Popkin BM, Du S. 'Dynamics of the nutrition transition toward the animal foods sector in China and its implications: A worried perspective' *Journal of Nutrition* 2003; 133: 3898S–3906S; McMichael AJ, Powles JW, Butler CD, Uauy R. 'Food, livestock production, energy, climate change and health' *Lancet* 2007; 370: 1253–1263.
6. Food and Agriculture Organization of the United States. *FAOSTAT*, 2009. Faostat.fao.org, date accessed December 4, 2010.
7. Sørensen JT, Edwards S, Noordhuizen J, Gunnarsson S. 'Animal production systems in the industrialised world' *Rev Sci Tech.* 2006; 25:493–503.
8. Akhtar AZ, Greger M, Ferdowsian H, Frank E. 'Health professionals' roles in animal agriculture, climate change, and human health' *American Journal of Preventive Medicine* 2009; 36: 182–187.
9. Jonathan Safran Foer, *Eating Animals* (New York: Little, Brown & Company) 2009.
10. Pearson et al., 2005.

11. Delgado C, Rosegrant M, Steinfeld H, Ehui S, Courbois C. 'Livestock to 2020, the next food revolution. *Food, agriculture, and the environment discussion paper 28*' (Washington, DC: International Food Policy Research Institute/FAO/International Livestock Research Institute) 2009.
12. Leibler JH, Otte J, Roland-Holst D et al. 'Industrial food animal production and global health risks: Exploring the ecosystems and economics of avian influenza' *EcoHealth* 2009; 6: 58–70.
13. Akhtar et al., 2009.
14. Leibler et al., 2009.
15. Pearson et al., 2005.
16. Weiss R. 'Report target costs of factory farming' *Washington Post* August 30, 2008.
17. Sørensen et al., 2006.
18. Greger M (a). 'The human/animal interface: Emergence and resurgence of zoonotic infectious diseases' *Critical Reviews in Microbiology* 2007; 33: 243–299.
19. Kestin SC, Knowles TG, Tinch AE, Gregory NG. 'Prevalence of leg weakness in broiler chick and its relationship with genotype' *The Veterinary Record* 1992; 131: 190–194; Duncan IJH. 'Welfare problems of meat-type chickens' Farmed Animal Well-Being Conference at the University of California-Davis, June 28–29, 2001; Knowles TG, Kestin SC, Haslam SM et al. 'Leg disorders in broiler chickens: Prevalence, risk factors and prevention' *Public Library of Science ONE* 2008; 3 (2): e1545. doi:10.1371/journal.pone.0001545.
20. Grandin T, Johnson C. *Animals in Translation* (New York: Scribner) 2005, pp. 270–271.
21. United Egg Producers. 'Housing, space, feed and water' 2009. www.uepcertified.com, date accessed December 11, 2010; Greger (a), 2007; Fraser D, Mench J, Millman S. 'Farm animals and their welfare in 2000' in Salem DJ, Row AN (ed.), *State of the Animals* (Washington, DC: Humane Society Press) 2001, pp. 87–99.
22. United Egg Producers. *United Egg Producers Animal Husbandry Guidelines for U.S Egg Laying Flocks, 2008 Edition* (Alpharetta, GA: United Egg Producers) 2008, p. 11. www.uepcertified.com, date accessed July 7, 2009.
23. Leibler et al., 2009.
24. Humane Society of the United State (HSUS). 'An HSUS Report: The welfare of animals in the egg industry'. www.humanesociety.org, date accessed November 9, 2010.
25. Compassion in World Farming. 'Meat chickens'. www.ciwf.org.uk, date accessed December 2, 2010.
26. Humane Society of the United States (HSUS). 'An HSUS report: The welfare of animals in the meat, egg and dairy industries'. www.humanesociety.org, date accessed November 9, 2010.
27. Nierenberg D. 'Happier meals: Rethinking the global meat industry' Worldwatch Institute, paper 171, 2005.
28. Sørensen et al., 2006.
29. Greger (a), 2007.
30. HSUS, 'The welfare of animals in the meat...'; Greger (a), 2007.

31. HSUS, 'The welfare of animals in the meat...'.
32. Ibid.; Sørensen et al., 2006.
33. The Veal Farm. 'Frequently asked questions'. www.vealfarm.com, date accessed December 3, 2010; Bentham J. 'Veal, without the cruelty' *Guardian* September 5, 2007. www.guardian.co.uk, date accessed December 3, 2010.
34. Weiss, 2008.
35. Greger (a), 2007; Sørensen et al., 2006.
36. Greger M (b). 'The long haul: Risks associated with livestock transport' *Biosecurity and Bioterrorism: Biodefense Strategy, Practice and Science* 2007; 5: 301–311.
37. Food and Agriculture Organization of the United Nations. *Guidelines for Humane Handling, Transport and Slaughter of Livestock* (Bangkok: FAO) 2001. www.fao.org/, date accessed December 9, 2010.
38. HSUS, 'The welfare of animals in the meat...'.
39. The PEW Charitable Trust. 'Industrial farm animal production, antimicrobial resistance and human health' 2008. www.pewtrusts.org, date accessed December 1, 2010.
40. Gilchrist MJ, Greko C, Wallinga DB, Beran GW, Riley DG, Thorne PS. 'The potential role of concentrated animal feeding operations in infectious disease epidemics and antibiotic resistance' *Environmental Health Perspectives* 2007; 115: 313–316.
41. Ibid.; Leibler et al., 2009.
42. Humane Society of the United States (HSUS). 'Food safety and cage egg production'. www.hsus.org, date accessed November 10, 2010.
43. Mulder R. 'Impact of transport and related stresses on the incidence and extent of human pathogens in pigmeat and poultry' *Journal of Food Safety* 1995; 15: 239–246.
44. Greger (a), 2007; Tuyttens FAM. 'The importance of straw for pigs and cattle welfare: A review' *Applied Animal Behaviour Science* 2005; 92: 261–282; Mulder, 1995.
45. Tuyttens, 2005.
46. Greger (b), 2007; Tuyttens, 2005; Mulder, 1995.
47. Whyte P, Collins JD, McGill K, Monahan C, O'Mahony H. 'The effect of transportation stress on excretion rates of Campylobacters in market-age broilers' *Poultry Science* 2001; 80: 817–820; Marg H, Scholz HC, Arnold T, Rosler U, Hensel A. 'Influence of long-time transportation stress on reactivation of *Salmonella typhimurium* DT 104 in experimentally infected pigs' *Berliner und Munchener Tierarztliche Wochenschrift* 2001; 114: 385–388.
48. Barham AR, Barham BL, Johnson AK et al. 'Effects of the transportation of beef cattle from the feedyard to the packing plant on prevalence levels of *Escherichia coli* O157 and *Salmonella* spp' *Journal of Food Protection* 2002; 65: 280–283.
49. Ibid.
50. Arthur TM, Bosilevac JM, Brichta-Harhay DM et al. 'Transportation and lairage environment effects on prevalence, numbers, and diversity of *Escherichia coli* O157:H7 on hides and carcasses of beef cattle at processing' *Journal of Food Protection* 2007; 70: 280–286.
51. Shackelford AD. 'Modifications of processing methods to control *Salmonella*' *Poultry Science* 1988; 67: 933–935.

52. Arthur et al., 2007; Mulder, 1995.
53. Cole DJ, Hill VR, Humenik FJ, Sobsey MD. 'Health, safety, and environmental concerns of farm animal waste' *Occupational Medicine: State of the Arts Reviews* 1999; 14: 423–428; Mitloehner FM, Schenker MB. 'Environmental exposure and health effects from concentrated animal feeding operations' *Epidemiology* 2007; 18: 309–311; Just N, Duchaine C, Singh B. 'An aerobiological perspective of dust in caged-house and floor housed poultry operations' *Journal of Occupational Medicine and Toxicology* 2009; 4: 13.
54. Mitloehner and Schenker, 2007; Thu KM. 'Public health concerns for neighbors of large-scale swine operations' *Journal of Agricultural Safety and Health* 2002; 8: 175–184; Greger M, Koneswaran G. 'The public health impacts of concentrated animal feeding operations on local communities' *Fam Community Health* 2010; 33: 11–20; Just et al., 2009.
55. Thu, 2002.
56. European Commission. 'The welfare of chickens kept for meat production (broilers), March 21' Scientific Committee on Animal Health and Animal Welfare (SCAHAW) 2000. ec.europa.eu, date accessed November 10, 2010.
57. Scientific American. 'Our sick farms, our infected food' *Scientific American* March 12, 2009. www.scientificamerican.com, date accessed February 13, 2010; Bennett C. 'Crohn's disease, sick cows and contaminated milk' *World Net Daily* October 1, 2004. www.wnd.com, date accessed December 3, 2010.
58. Pearson et al., 2005.
59. Centers for Disease Control and Prevention (CDC). 'Investigation update: Multistate outbreak of human *Salmonella enteritidis* infections associated with shell eggs' December 2, 2010. www.cdc.gov, date accessed June 26, 2011.
60. Mead PS, Slutsker L, Dietz V et al. 'Food-related illness and death in the United States' *Emerging Infectious Diseases* 1999; 5: 607–625; Chittick P, Sulka A, Tauxe RC, Fry AM. 'A summary of national reports of foodborne outbreaks of *Salmonella* Heidelberg infections in the United States: Clues for disease prevention' *Journal of Food Protection* 2006; 69; 1150–1153.
61. Leirisalo-Repo M, Helenius P, Hannu T et al. 'Long-term prognosis of reactive salmonella arthritis' *Annals of the Rheumatic Diseases* 1997; 56: 516–520.
62. Food and Drug Administration (FDA). '*Salmonella enteritidis* outbreak in shell eggs' October 18, 2010. www.fda.gov/food/newsevents/whatsnewin food/ucm222684.htm#, date accessed November 10, 2010.
63. Marsh B. 'A hen's space to roost' *New York Times* August 14, 2010.
64. Van Hoorebeke S , Van Immerseel FV, Schulz J et al. (a). 'Determination of the within and between flock prevalence and identification of risk factors for *Salmonella* infections in laying hen flocks housed in conventional and alternative systems' *Preventive Veterinary medicine* 2010; 94; 94–100; Methner U, Diller R, Reiche R, Bohland K. 'Occurrence of salmonellae in laying hens in different housing systems and inferences for control' *Berl Munch Tieraztl Wochenschr* 2006; 119: 467–473; De Vylder J, Van Hoorebeke S, Ducatelle R et al. 'Effect of the housing system on shedding and colonization of gut and internal organs of laying hens with *Salmonella enteritidis*' *Poultry Science* 2009; 88: 2491–2495; Snow LC, Davies RH, Christiansen KH et al. 'Survey of the prevalence of *Salmonella* species on commercial

laying farms in the United Kingdom' *The Veterinary Record* 2007; 161: 471–476; Huneau-Salaün A, Chemaly M, Le Bouquin S et al. 'Risk factors for *Salmonella enterica* subsp. *enterica* contamination in 519 French laying hen flocks at the end of the laying period' *Preventive Veterinary Medicine* 2009; 89: 51–58; Namata H, Méroc E, Aerts et al. '*Salmonella* in Belgian laying hens: An identification of risk factors' *Preventive Veterinary Medicine* 2008; 83: 323–336; Mahé A, Bougeard S, Huneau-Salaün A et al. 'Bayesian estimation of flock-level sensitivity of detection of *Salmonella* spp., Enteriditis and Typhimurium according to the sampling procedure in French laying-hen houses' *Preventive Veterinary Medicine* 2008; 84: 11–26; Snow LC, Davies RH, Christiansen KH et al. 'Investigation of risk factors for *Salmonella* on commercial egg-laying farms in Great Britain, 2004–2005' *The Veterinary Record* 2010; 166: 579–586.

65. Namata et al., 2008; Davies R, Breslin M. 'Observations on *Salmonella* contamination of commercial laying farms before and after cleaning and disinfection' *The Veterinary Record* 2003; 152: 283–287.

66. Ibid.; Olsen AR, Hammack TS. 'Isolation of *Salmonella* spp. from the housefly, *Musca domestica* L., and the dump fly *Hydrotaea aenescens* (Wiedemann) (Diptera: Muscidae), at caged-layer houses' *Journal of Food Protection* 2000; 63: 958–960; Winpisinger KA, Ferketich AK, Berry RL, Moeschberger ML. 'Spread of *Musca domestica* (Diptera: muscidae), from two caged layer facilities to neighboring residences in rural Ohio' *Journal of Medical Entomology* 2005; 42: 732–738; Garber L, Smeltzer M, Fedorka-Cray P, Ladely S, Ferris K. '*Salmonella enterica* subtype serotype Enteritidis in table egg layer house environments and in mice in U.S. layer houses and associated risk factors' *Avian Diseases* 2003; 47: 134–142; Van Hoorebeke S, Van Immerseel F, Haesebrouck F, Ducatelle R, Dewulf J (b). 'The influence of the housing system on *Salmonella* infections in laying hens: A review' *Zoonoses and Public Health* September 28, 2010. doi: 10.1111/j.1863–2378.2010.01372.x. (epub ahead of print).

67. Van Hoorebeke et al. (b), 2010; Eckholm E. 'Egg industry faces new scrutiny after outbreak' *New York Times*, August 23, 2010. http://www.nytimes.com/2010/08/24/us/24eggs.html, date accessed November 20, 2010.

68. European Food Safety Authority. 'Report of the task force on zoonoses data collection on the analysis of the baseline study on the prevalence of *Salmonella* in holdings of laying hen flocks of *Gallus gallus*' *The EFSA Journal* 97, 2007. http://www.efsa.europa.eu/EFSA/efsa_locale-1178620753812_1178620761896.htm, date accessed November 21, 2010.

69. Centers for Disease Control and Prevention (CDC). 'Preliminary FoodNet data on the incidence of infection with pathogens transmitted commonly through food—10 states, United States, 2005' *Morbidity and Mortality Weekly Report* 2006; 55: 392–395.

70. Nakamura M, Nagamine N, Takahashi T et al. 'Horizontal transmission of *Salmonella enteritidis* and effect of stress on shedding in laying hens' *Avian Diseases* 1994; 38: 282–288.

71. Just et al., 2009.

72. Sarwari AR, Magder LS, Levine P et al. 'Serotype distribution of *Salmonella* isolates from food animals after slaughter differs from that of isolates found in humans' *Journal of Infectious Diseases* 2001; 183: 1295–1299.

73. Poppe C, Irwin RJ, Forsberg CM, Clarke RC, Oggel J. 'The prevalence of *Salmonella enteritidis* and other *Salmonella* spp. among Canadian registered commercial layer flocks' *Epidemiology and Infection* 1991; 106: 259–270; Poppe C, Irwin RJ, Messier S, Finley GG, Oggel J. 'The prevalence of *Salmonella enteritidis* and other *Salmonella* spp. among Canadian registered commercial broiler flocks' *Epidemiology and Infection* 1991; 107: 201–211.

74. Maes D, Deluyker H, Verdonck M et al.'Herd factors associated with the seroprevalences of four major respiratory pathogens in slaughter pigs from farrow-to-finish pig herds' *Veterinary Research* 2000; 31: 313–327; Jones PW, Colins P, Brown GT, Aitken MM. '*Salmonella saint-paul* infection in two dairy herds' *Journal of Hygiene* 1983; 91: 243–257; Lanada EB, Morris RS, Jackson R, Fenwick SG. 'Prevalence of *Yersinia* species in goat flocks' *Australian Veterinary Journal* 2005; 83: 563–566; Salman MD, Meyer ME. 'Epidemiology of bovine brucellosis in the Mexicali Valley, Mexico: Literature review of disease-associated factors' *American Journal of Veterinary Research* 1984; 45: 1557–1560; Atwill ER, Johnson EM, Pereira MG. 'Association of herd composition, stocking rate, and duration of calving season with fecal shedding of *Cryptosporidium parvum* oocysts in beef herds' *Journal of the American Veterinary Medical Association* 1999; 215: 1833–1838; Stacey KF, Parsons DJ, Christiansen KH, Burton CH. 'Assessing the effect of interventions on the risk of cattle and sheep carrying *Escherichia coli* O157:H7 to the abbatoir using a stochastic model' *Preventive Veterinary Medicine* 2007; 79: 32–45.

75. Withers MR, Correa MT, Morrow M et al. 'Antibody levels to hepatitis E virus in North Carolina swine workers, non-swine workers, swine and murids' *American Journal of Tropical Medicine and Hygiene* 2002; 66: 384–388; Olsen B, Axelsson-Olsson D, Thelin A, Weiland O. 'Unexpected high prevalence of IgG-antibodies to hepatitis E virus in Swedish pig farmers and controls' *Scandinavian Journal of Infectious Diseases* 2006; 38: 55–58; Olsen CW, Brammer L, Easterday BC et al. 'Serologic evidence of H1swine influenza virus in swine farm residents and employees' *Emerging Infectious Diseases* 2002; 8: 814–819; Fey PD, Safranek TJ, Rupp ME et al. 'Ceftriaxone-resistant salmonella infection acquired by a child from cattle' *New England Journal of Medicine* 2000; 342: 1242–1249; van den Bogaard AE, London N, Criessen C, Stobberingh EE. 'Antibiotic resistance of faecal *Escherichia coli* in poultry, poultry farmers and poultry slaughterers' *Journal of Antimicrobial Chemotherapy* 2001; 47: 763–771; van den Bogaard AE, Willems R, London N, Top J, Stobberingh EE. 'Antibiotic resistance of faecal enterococci in poultry, poultry farmers and poultry slaughterers' *Journal of Antimicrobial Chemotherapy* 2002; 49: 497–505; Gray GC, Baker WS. 'The importance of including swine and poultry workers in influenza vaccination programs' *Clinical Pharmacology & Therapeutics* 2007; 82: 638–641.

76. Myers KP, Olsen CW, Setterquist SF et al. 'Are swine workers in the United States at increased risk of infection with zoonotic influenza virus?' *Clinical Infectious Diseases* 2006; 42: 14–20; Cole D, Todd L, Wing S. 'Concentrated swine feeding operations and public health: A review of occupational and community health effects' *Environmental Health Perspectives* 2000; 108; 685–699.

77. Millner PD. 'Bioaerosols associated with animal production operations' *Bioresource Technology* 2009; 100: 5379–5385.
78. Green CF, Gibbs SG, Tarwater PM, Mota LC, Scarpino PV. 'Bacterial plume emanating from the air surrounding swine confinement operations' *Journal of Occupational Environmental Hygiene* 2006; 3: 9–15; Millner, 2009.
79. Green et al., 2006.
80. Booth B. 'Microbes in the air near swine farms—More or less' *Environmental Science and Technology* 2008. doi 10.1021/es802620k.
81. Gibbs SG, Green CF, Tarwater PM, Scarpino PV. 'Airborne antibiotic resistant and nonresistant bacteria and fungi recovered from two swine herd confined animal feeding operations' *Journal of Occupational Environmental Hygiene* 2004; 1: 699–706; Gibbs SG, Green CF, Tarwater PM et al. 'Isolation of antibiotic-resistant bacteria from the air plume downwind of a swine confined or concentrated feeding operation' *Environmental Health Perspective* 2006; 114: 1032–1037; Ko G, Simmons III OD, Likirdopulos CA, Worley-Davis L, Williams M, Sobsey MD. 'Investigation of bioaerosols released from swine farms using conventional and alternative waste treatment and management technologies' *Environmental Science and Technology* 2008; 42: 8849–8857.
82. Waltner-Toews D, Lang T. 'A new conceptual base for food and agricultural policy: The emerging model of links between agriculture, food, health, environment and society' *Global Change and Human Health* 2000; 1, 116–130.
83. Tauxe RV. 'Emerging foodborne pathogens' *International Journal of Food Microbiology* 2002; 78: 31–41; Mead et al., 1999.
84. Delgado et al., 1999.
85. Adams M, Motarjemi Y. *Basic Food Safety for Health Workers* (Geneva: WHO Press) 1999; McMichael AJ, Haines A, Slooff R, Kovats S (eds), *Climate Change and Human Health* (Geneva: WHO, World Meteorological Organization, United States Environmental Program) 1996; Lederberg J, Shope RE, Oaks SC. *Emerging Infections: Microbial Threats to Health in the United States* (Washington, DC: National Academies Press) 1992, p. 15.
86. Johnson JR, Kuskowski MA, Smith K, O'Bryan TT, Tatini S. 'Antimicrobial-resistant and extraintestinal pathogenic *Escherichia coli* in retail foods' *Journal of Infectious Diseases* 2005; 191:1040–1049.
87. Centers for Disease Control (CDC). 'Foodborne illness'. www.cdc.gov, date accessed December 9, 2010.
88. McCarthy N, Giesecke J. 'Incidence of Guillan-Barre Syndrome following infection with *Campylobacter jejuni*' *American Journal of Epidemiology* 2001; 153: 610–614.
89. Spencer JL, Guan J. 'Public health implications related to spread of pathogens in manure from livestock and poultry operations' *Methods in Molecular Biology* 2004; 268: 503–515.
90. Duhigg C. 'Health ills abound as farm runoff fouls wells' *New York Times* September 18, 2009.
91. Martinez J, Dabert P, Barrington S, Burton C. 'Livestock waste treatment systems for environmental quality, food safety, and sustainability' *Bioresource Technology* 2009; 100: 5527–5536.
92. Leibler et al., 2009.

93. Gerba CP, Smith JE. 'Sources of pathogenic microorganisms and their fate during land application of wastes' *Journal of Environmental Quality* 2005; 34: 42–48; Ed Ayres, 'Will we still eat meat?' *Time Magazine* November 8, 1999.
94. Johns Hopkins Bloomberg School of Public Health. 'Public health project to address problems of animal production' September 9, 1999. www.jhsph. edu, date accessed December 4, 2010.
95. Graham JP, Leibler JH, Price LB et al. 'The animal-human interface and infectious disease in industrial food animal production: Rethinking biosecurity and biocontainment' *Public Health Reports* 2008; 123: 282–299.
96. Wing S, Freedman S, Band L. 'The potential impact of flooding on confined animal feeding operations in Eastern North Carolina' *Environmental Health Perspectives* 2002; 110: 387–391.
97. Spencer and Guan, 2004; Leibler et al., 2009.
98. Gerba et al., 2005.
99. Jenkins, M. 'Persistence and transport of pathogens from animal agriculture in soil and water' *Water Environment Research* 2009; 347–368.
100. Guan Ty, Holley RA. 'Pathogen survival in swine manure environments and transmission of human enteric illness—A review' *Journal of Environmental Quality* 2003; 32: 383–392; Alberta P. 'Industry statistics' *Alberta Pork*. www. albertapork.com, date accessed December 9, 2010.
101. Xiao L, Moore JE, Ukoh U et al. 'Prevalence and identity of *Cryptosporidium* spp. in pig slurry' *Applied and Environmental Microbiology* 2006; 72: 4461–4463.
102. Guan and Holley, 2003.
103. Gerba et al., 2005.
104. Spencer and Guan, 2004.
105. Riemann H, Himathongkham S, Willoughby D, Tarbell R, Breitmeyer R. 'A survey for *Salmonella* by drag swabbing manure piles in California egg ranches' *Avian Diseases* 1998; 42: 67–71.
106. Guan and Holley, 2003; Himathongkham S, Bahari S, Riemann H, Cliver D. 'Survival of *Escherichia coli* O157:H7 and *Salmonella typhimurium* in cow manure and cow manure slurry' *FEMS Microbiology Letters* 1999; 178: 251–257.
107. Jenkins, 2009.
108. Guan and Holley, 2003.
109. Ibid.
110. Ibid.
111. Pell AN. 'Manure and microbes: Public and animal health problem?' *Journal of Dairy Science* 1997; 80: 2673–2681.
112. Tauxe, 2002; Chemaly M, Toquin M-T, le Nôtre Y, Fravalo P. 'Prevalence of *Listeria monocytogenes* in poultry production in France' *Journal of Food Protection* 2008; 71: 1996–2000.
113. Mead et al., 1999.
114. MSN. 'Source of tainted spinach finally pinpointed' *MSNBC.com* March 23, 2007. www.msnbc.com, date accessed January 9, 2008.
115. Guo X, Chen J, Brackett RE, Beauchat LR. 'Survival of salmonellae on and in tomato plants from the time of inoculation at flowering and early stages of fruit development through fruit ripening' *Applied and Environmental Microbiology* 2001; 67: 4760–4764.

116. Michino H, Araki K, Minami S et al. 'Massive outbreak of *Escherichia coli* O157:H7 infection in school children in Sakai City Japan, associated with consumption of white radish sprouts' *American Journal of Epidemiology* 1999; 150: 787–796.
117. Spencer and Guan, 2004.
118. Ibid.
119. Greger (a), 2007.
120. Valcour JE, Michel P, McEwen SA, Wilson JB. 'Associations between indicators of livestock farming intensity and incidence of human Shiga toxin-producing *Escherichia coli* infection' *Emerging Infectious Diseases* 2002; 8: 252–257; Kistemann T, Zimmer S, Vagsholm I, Andersson Y. 'GIS-supported investigation of human EHEC and cattle VTEC O157 infections in Sweden: Geographical distribution, spatial variation and possible risk factors' *Epidemiology and Infection* 2004; 132: 495–505; Haus-Cheymol R, Espie E, Che D et al. 'Association between indicators of cattle density and incidence of paediatric haemolytic-ureamic syndrome (HUS) in children under 15 years of age in France between 1996 and 2001: An ecological study' *Epidemiology and Infection* 2005; 134: 1–7; Michel P, Wilson JB, Martin SW et al. 'Temporal and geographical distributions of reported cases of *Escherichia coli* O157:H7 infection in Ontario' *Epidemiology and Infection* 1999; 122: 193–200; Frank C, Kapfhammer S, Werber D, Stark K, Held L. 'Cattle density and Shiga toxin-producing *Escherichia coli* infection in Germany: increased risk for most but not all serogroups' *Vector Borne and Zoonotic Diseases* 2008; 8: 635–643.
121. Greger (a), 2007.
122. PEW Charitable Trust, 2008.
123. Tomasz A. 'Multiple-antibiotic-resistant pathogenic bacteria—A report on the Rockefeller University Workshop' *New England Journal of Medicine* 1994; 330: 1247–1251.
124. Peeples L. 'Bill proposed to limit livestock antibiotics to prevent the rise of resistant germs' *Scientific American* July 14, 2009. www.scientific American.com/blog, date accessed November 13, 2010.
125. PEW Charitable Trust, 2008.
126. Aarestrup FM, Duran CO, Burch DGS. 'Antimicrobial resistance in swine production' *Animal Health Research Reviews* 2008; 9: 135–148; Gilchrist et al., 2007; Hughes P, Heritage J. 'Antibiotic growth-promotors in food animals' in *Assessing Quality and Safety of Animal Feeds* (Rome: Food and Agriculture Organization of the United Nations) 2004.
127. Hughes and Heritage, 2004.
128. Office of Technology Assessment. *Drugs in Livestock Feed. Volume 1: Technical Report* (Washington, DC: US Government Printing Office) 1979, p. 41. www.princeton.edu, date accessed November 21, 2010.
129. Hughes and Heritage, 2004.
130. *Scientific American*, 2009.
131. Akhtar et al., 2009.
132. Mellon M, Benbrook C, Benbrook KL. *Hogging It! Estimates of Antimicrobial Abuse in Livestock* (Cambridge, MA: Union of Concerned Scientists) 2001.

133. World Health Organization (WHO). 'Use of antimicrobials outside human medicine and resultant antimicrobial resistance in humans' *WHO* 2002. apps.who.int/inf-fs/en, date accessed November 21, 2010.

134. Pew Charitable Trust, 2008.

135. Perkin RM, Swift JD, Newton DA, Anas NG. (eds) *Hospital Medicine Textbook of Inpatient Medicine*, 2nd edn. (Philadelphia, PA: Lippincott, Williams, & Wilkins) 2008, p. 450.

136. Pecquet J. 'Animal drug industry questions antibiotic resistance claims' *The Hill* July 20, 2010. http://thehill.com/blogs, date accessed November 3, 2010; Murphree J. 'Continuing the conversation on animal agriculture's antibiotic use: The poultry industry' *Arizona Farm Bureau*, 2008. www.azfb. org, date accessed December 11, 2010.

137. Smith TC, Male MJ, Harper AL et al. (a). 'Methicillin-resistant *Staphylococcus aureus* (MRSA) strain ST398 is present in Midwestern U.S. swine and swine workers' *Public Library of Science One* 2009; 4 (1): e4258. doi:10.1371/journal.pone.0004258.

138. Kuehn BM. 'Antibiotic-resistant "superbugs" may be transmitted from animals to humans' *Journal of the American Medical Association* 2007; 298: 2125–2126; Levy SB, Fitzgerald GB, Macone AB. 'Changes in intestinal flora of farm personnel after introduction of a tetracycline-supplemented feed on a farm' *New England Journal of Medicine* 1976; 295: 538–588.

139. Gibbs et al., 2006; Gibbs et al., 2004; Campagnolo ER, Johnson KR, Karpati A et al. 'Antimicrobial residues in animal waste and water resources proximal to large-scale swine and poultry feeding operations' *Science of the Total Environment* 2002; 299: 89–95; Price LB, Graham JP, Lackey LG et al. 'Elevated risk of carrying gentamicin-resistant *Escherichia coli* among U.S. poultry workers' *Environmental Health Perspectives* 2007; 115: 1738–1742; Yang H, Dettman B, Beam J, Mix C, Jiang X. 'Occurrence of ceftriaxone-resistant commensal bacteria on a dairy farm and a poultry farm' *Canadian Journal of* Microbiology 2006; 52: 942–950; Siegel D, Huber WG, Enloe F. 'Continuous non-therapeutic use of antibacterial drugs in feed and drug resistance of the gram-negative enteric florae of food-producing animals' *Antimicrobial Agents and Chemotherapy* 1974; 6: 697–701; Nógrády N, Kardos G, Bistyák A et al. 'Prevalence and characterization of *Salmonella infantis* isolates originating from different points of the broiler chicken-human food chain in Hungary' *International Journal of Food Microbiology* 2008; 127: 162–167; Amyes SG. 'Trimethoprim resistance in commensal bacteria isolated from farm animals' *Epidemiology and Infection* 1987; 98: 87–96; Ghosh S, LaPara TM. 'The effects of subtherapeutic antibiotic use in farm animals on the proliferation and persistence of antibiotic resistance among soil bacteria' *Intenational Society for Microbial Ecology Journal* 2007; 1:191–203; Welch B, Forsberg CW. 'Chlortetracycline and sulfonamide resistance of fecal bacteria in swine receiving medicated feed' *Canadian Journal of Microbiology* 1979; 25: 789–792; Smith HW, Lovell MA. '*Escherichia coli* resistant to tetracyclines and to other antibiotics in the faeces of U.K. chickens and pigs in 1980' *Journal of Hygiene (London)* 1981; 87: 477–483; Sayah RS, Kaneene JB, Johnson Y, Miller R. 'Patterns of antimicrobial resistance observed in *Escherichia coli* isolates obtained from domestic- and wild-animal fecal samples, human

septage, and surface water' *Applied and Environmental Microbiology* 2005; 71: 1394–1404.

140. Graham JP, Price LB, Evans SL, Graczyk TK, Silbergeld EK. 'Antibiotic resistant enterococci and staphylococci isolated from flies collected near confined poultry feeding operations' *Science of the Total Environment* 2009; 407: 2701–2710.
141. Campagnolo et al., 2002.
142. Akwar TH, Poppe C, Wilson J et al. 'Risk factors for antimicrobial resistance among fecal *Escherichia coli* from residents on forty-three swine farms' *Microbial Drug Resistance* 2007; 13: 69–76.
143. Chapin A, Rule A, Gibson K, Buckley T, Schwab K. 'Airborne multi-drug resistant bacteria isolated from a concentrated swine feeding operation' *Environmental Health Perspective* 2005; 113: 137–142.
144. Gibbs et al., 2006.
145. Johnson JR, Kuskowski MA, Smith K, O'Bryan TT, Tatini S. 'Antimicrobial-resistant and extraintestinal pathogenic *Escherichia coli* in retail foods' *Journal of Infectious Disease* 2005; 191: 1040–1049; Manie T, Khan S, Brözel VS, Veith WJ, Gouws PA. 'Antimicrobial resistance of bacteria isolated from slaughtered and retail chickens in South Africa' *Letters in Applied Microbiology* 1998; 26: 253–258; Nógrády et al., 2008; Holmberg SD, Osterholm MT, Senger KA, Cohen ML. 'Drug-resistant salmonella from animals fed antimicrobials' *New England Journal of Medicine* 1984; 311: 617–622; Spika JS, Wasterman SH, Soo Hoo GW et al. 'Chloramphenicol-resistant *Salmonella newport* traced through hamburger to dairy farms' *New England Journal of Medicine* 1987; 316: 565–570; Smith KE, Besser JM, Hedberg CW et al. 'Quinolone-resistant *Campylobacter jejuni* infections in Minnesota, 1992–1998' *New England Journal of Medicine* 1999; 340: 1525–1532; Sanotra et al., 2001; Arnold S, Gassner B, Giger T, Zwahlen R. 'Banning antimicrobial growth promoters in feedstuffs does not result in increased therapeutic use of antibiotics in medicated feed in pig farming' *Pharmacoepidemiology and Drug Safety* 2004; 13: 323–331.
146. NARMS. *National Antimicrobial Resistance Monitoring System Animal Isolates.* www.ars.usda.gov, date accessed December 3, 2010.
147. Sapkota AR, Lefferts LY, McKenzie S, Walker P. 'What do we feed to food-production animals? A review of animal feed ingredients and their potential impacts on human health' *Environmental Health Perspectives* 2007; 115: 663–670; Smith et al., 1999.
148. Smith et al., 1999.
149. Kaufman M. 'Ending battle with FDA, Bayer withdraws poultry antibiotic' *Washington Post* September 9, 2005.
150. 'Antibiotic resistance and the use of antibiotics in animal agriculture' Statement of Joshua M. Sharfstein, Principle Deputy Comissioner. Food and Drug Administration before Subcommitte on Health, July 14, 2010. www.fda.gov, date accessed December 9, 2010.
151. Loglisci RF. 'On antibiotic resistance in food animals' *Food Safety News* July 8, 2010. www.foodsafetynews.com, date accessed December 10, 2010.
152. Treanor J. 'Influenza vaccine—Outmaneuvering antigenic shift and drift' *New England Journal of Medicine* 2004; 350: 218–220.

153. Skeik N, Jabr FI. 'Influenza viruses and the evolution of avian influenza virus H5N1' *International Journal of Infectious Diseases* 2008; 12: 233–238.
154. Webster RG, Bean WJ, Gorman OT, Chambers TM, Kawaoka Y. 'Evolution and ecology of influenza A viruses' *Microbiological Reviews* 1992; 56: 152–179.
155. World Health Organization (WHO). 'Statements of 2009. Influenza A (H1N1)' April 29, 2009. www.who.int/mediacentre/news/statements/2009/en, date accessed November 17, 2010.
156. Taubenberger JK, Morens DM. '1918 influenza: The mother of all pandemics' *Emerging Infectious Diseases* 2006; 12: 15–22.
157. Neumann G, Noda T, Kawaoka Y. 'Emergence and pandemic potential of swine-origin H1N1 influenza virus' *Nature* 2009; 459: 931–939; Taubenberger and Morens, 2006.
158. Lallanilla M. 'Spanish flu of 1918. Could it happen again?' *ABC News*, October 5, 2005. abcnews.go.com, date accessed November 4, 2010.
159. Eagles D, Sireger ES, Dung DH,et al. 'H5N1 highly pathogenic avian influenza in Southeast Asia' *Rev sci tech off int epiz* 2009; 28: 341–348; World Health Organization (WHO). 'H5N1 avian influenza: Timeline of major events, May 2, 2011.' www.who.int, date accessed November 19, 2010.
160. Webster et al., 1992.
161. Webby RJ, Webster RG. 'Emergence of influenza A viruses' *Philosophical Transactions of the Royal Society of London B Biological Sciences* 2001; 356: 1817–1828.
162. Yee KS, Carpenter TE, Cardona CJ. 'Epidemiology of H5N1 avian influenza' *Comparative Immunology, Microbiology and Infectious Diseases* 2009; 32: 325–340.
163. Ibid.; Peiris JSM, de Jong MD, Guan Y. 'Avian influenza virus (H5N1): A threat to human health' *Clinical Microbiology Reviews* 2007; 20: 243–267.
164. Capua I, Alexander DJ. 'Animal and human health implications of avian influenza infections' *Bioscience Reports* 2007; 27: 359–372.
165. Peiris et al., 2007.
166. Ibid.
167. Mackenzie D. 'Introduction: Bird flu' *New Scientist* September 4, 2006.
168. Peiris et al., 2007; Tiensin T, Sayeem S, Ahmed SU et al. 'Ecologic risk factor investigation of clusters of Avian Influenza A (H5N1) virus infection in Thailand' *Journal of Infectious Diseases* 2009; 199: 1735–1743.
169. WHO, 'H5N1 avian influenza: Timeline', 2011.
170. World Health Organization (WHO). 'Avian Influenza ("bird flu")' February 2006. www.who.int, date accessed November 10, 2010.
171. WHO, 'H5N1 avian influenza: Timeline' 2011; World Health Organization (WHO). 'Cumulative number of confirmed human cases of Avian Influenza A (H5N1) reported to WHO'. www.who.int, date accessed June 20, 2011.
172. Yee et al., 2009; Chen J-M, Chen J-W, Dai J-J, Sun Y-X. 'A survey of human cases of H5N1 avian influenza reported by the WHO before June 2006 for infection control' *American Journal of Infection Control* 2007; 35:351–353; WHO, 'H5N1 avian influenza: Timeline' 2011; WHO, 'Cumulative number', 2009.
173. Peiris et al., 2007.

174. World Health Organization (WHO). 'Current phase of alert in the WHO global influenza preparedness plan (avian influenza H5N1)'. www.who.int/, date accessed November 20, 2010; Martinot A, Thomas J, Thiermann A, Dasgupta N. 'Prevention and control of avian influenza: The need for a paradigm shift in pandemic influenza preparedness' *The Veterinary Record* 2007; 160: 343–345; WHO, 'H5N1 avian influenza: Timeline', 2011; Peiris et al., 2007.
175. Iwami S, Takeuchi Y, Liu X. 'Avian flu pandemic: Can we prevent it?' *Journal of Theoretical* Biology 2009; 257: 181–190; Babakir-Mina M, Balestra E, Perno CF, Aquaro S. 'Influenza virus A (H5N1): A pandemic risk?' *New Microbiologica* 2007; 30: 65–77; WHO, 'Statements of 2009', 2009.
176. Capua and Alexander, 2007.
177. WHO, 'Avian influenza', 2006.
178. Capua and Alexander, 2007.
179. Peiris et al., 2007; Capua and Alexander, 2007; WHO, 'Avian influenza', 2006.
180. Peiris et al., 2007; Bavinck V, Bouma A, van Boven M et al. 'The role of backyard flocks in the epidemic of highly pathogenic avian influenza virus (H7N7) in the Netherlands in 2003' *Preventive Veterinary Medicine* 2009; 88: 247–254.
181. Babakir-Mina et al., 2007.
182. Iwami et al., 2009; Mackenzie, 2006.
183. Centers for Disease Control and Prevention (CDC). 'Past avian influenza outbreaks'. http://www.cdc.gov/flu/avian/outbreaks/past.htm, date accessed November 21, 2010; Capua and Alexander, 2007.
184. Alexander DJ, Brown IH. 'History of highly pathogenic avian influenza' *Rev Sci tech Off int Epiz* 2009; 28: 19–38.
185. Graham et al., 2008.
186. Leibler et al., 2009.
187. Graham et al., 2008.
188. Ibid.; Leibler et al., 2009.
189. Graham et al., 2008.
190. Bavinck et al., 2009.
191. Busani L, Valsecchi MG, Rossi E et al. 'Risk factors for highly pathogenic H7N1 avian influenza virus infection in poultry during the 1999–2000 epidemic in Italy' *The Veterinary Journal* 2009; 181: 171–177.
192. Leibler et al., 2009.
193. World Health Organizaion (WHO). 'Avian influenza: assessing the pandemic threat' 2005. www.who.int/, date accessed February 14, 2011.
194. Leibler et al., 2009.
195. Ibid.
196. Ibid.
197. Capua and Alexander, 2007.
198. WHO, 'Avian influenza', 2006.
199. Taubenberger and Morens, 2006.
200. Weingartl HM, Albrecht RA, Lager KM, et al. 'Experimental infection of pigs with the human 1918 pandemic influenza virus' *Journal of Virology* 2009; 83: 4287–4296.

201. Zhou NN, Senne DA, Landgraf JS, et al. 'Genetic reassortment of avian, swine, and human influenza A viruses in American pigs' *Journal of Virology* 1999; 73: 8851–8856; Weingartl et al., 2009; Brown IH. 'The epidemiology and evolution of influenza viruses in pigs' *Veterinary Microbiology* 2000; 74: 29–46; Pappaioanou M, Gramer M. 'Lessons from pandemic H1N1 2009 to improve prevention, detection, and response to influenza pandemics from a one health perspective' *Institute for Laboratory Animal Research Journal* 2010; 51: 268–280.
202. Pappaioanou and Gramer, 2010; Wuethrich B. 'Chasing the fickle swine flu' *Science* 2003; 299: 1502–1505.
203. Zhou NN, Senne DA, Landgraf JS et al. 'Emergence of H3N2 reassortment influenza A virus in North American pigs' *Veterinary Microbiology* 2000; 74: 47–58; Pappaioanou and Gramer, 2010; Wuethrich, 2003.
204. Wuethrich, 2003; Weingartl et al., 2009; Pappaioanou and Gramer, 2010; Wertheim JO. 'When pigs fly: The avian origin of "swine flu" *Environmental Microbiology* 2009; 11: 2191–2192.
205. The Washington Post. 'Little boy at the center of a viral storm' *Washington Post* April 29, 2009.
206. Webby RJ, Rossow K, Erickson G, Sims Y, Webster R. 'Multiple lineages of antigenically and genetically diverse influenza A virus co-circulate in the United States swine population' *Virus Research* 2004; 103: 67–73.
207. Wuethrich, 2003; Ma W, Lager KM, Vincent AL et al. 'The role of swine in the generation of novel influenza viruses' *Zoonoses Public Health* May 20, 2009 (epub ahead of print).
208. Sinha NK, Roy A, Das B, Das S, Basik S. 'Evolutionary complexities of swine flu H1N1 gene sequences of 2009' *Biochemical and Biophysical Research Communications* 2009; 390: 349–351; Wertheim, 2009; Girard MP, Tam JS, Assossou OM, Kieny MP. 'The 2009 A (H1N1) influenza virus pandemic: A review' *Vaccine* 2010; 28: 4895–4902; Garten RJ, Davis CT, Russell CA et al. 'Antigenic and genetic characteristics of swine-origin 2009 A (H1N1) influenza viruses circulating in humans' *Science* 2009; 325; 197–201; Smith GJD, Vijaykrishna D, Bahl J et al. (b). 'Origins and evolutionary genomics of the 2009 swine-origin H1N1 influenza A epidemic' *Nature* 2009; 459: 1122–1125.
209. Garten et al., 2009; Smith et al. (b), 2009.
210. Centers for Disease Control and Prevention (CDC). 'Origin of 2009 H1N1 flu (swine flu): questions and answers, November 25, 2009'. www.flu.gov, date accessed November 10, 2010.
211. Ibid.
212. Smith et al. (b), 2009.
213. Pappaioanou and Gramer, 2010.
214. Pearson et al., 2005; Wuethrich, 2003.
215. Saenz RA, Hethcote HW, Gray GC. 'Confined animal feeding operations as amplifiers of influenza' *Vector-Borne and Zoonotic Diseases* 2006; 6: 338–346; Myers et al., 2006; Olsen et al., 2002.
216. Butler D. 'Patchy pig monitoring may hide flu threat' *Nature* 2009; 459: 894–895.
217. Iwami et al., 2009; WHO, 'Avian influenza', 2006.

218. Capua and Alexander, 2007; Yee et al., 2009; Nidom CA, Takano R, Yamada S et al. 'Influenza A (H5N1) viruses from pigs, Indonesia' *Emerging Infectious Diseases* 2010; 16: 1515–1523.
219. Graham et al., 2008.
220. Nidom et al., 2010.
221. Mackenzie D. 'Bird flu jumps to pigs' *New Scientist* September 7, 2010. www. newscientist.com, date accessed November 10, 2010.
222. Cyranoski, 2009; Barrette et al., 2009.
223. Barrette et al., 2009; Cyranoski, 2009.
224. Normille, 2009.
225. Schmidt CW. 'Swine CAFOs and novel H1N1 flu: Separating facts from fears' *Environmental Health Perspectives* 2009; 117: A394–A401.
226. Wuethrich, 2003.
227. Leibler et al., 2009.
228. Graham et al., 2008.
229. Leibler et al., 2009; Graham et al., 2008.
230. Leibler et al., 2009.
231. Bull SA, Allen VM, Dominique G et al. 'Sources of *Campylobacter* spp. colonized housed broiler flocks during rearing' *Applied and Environmental Microbiology* 2006; 72: 645–652.
232. Leibler et al., 2009.
233. Ibid.
234. Saenz et al., 2006.
235. Peiris et al., 2007.
236. Taubenberger and Morens, 2006.
237. Yee et al., 2009.
238. Mackenzie D. 'Vietnam in U-turn over flu bird vaccination' *New Scientist* May 4, 2005. www.newscientist.com, date accessed November 21, 2010.
239. WHO, 'Avian influenza', 2006.
240. Mackenzie, 2005.
241. Zhou L, Liao Q, Dong L et al. 'Risk factors for human illness with avian influenza A (H5N1) virus infection in China' *Journal of Infectious Diseases* 2009; 199: 1726–1734; Mackenzie, 2005; WHO, 'H5N1 avian influenza: Timeline', 2011.
242. Hafez MH, Arafa A, Abdelwhab EM et al. 'Avian influenza H5N1 virus infections in vaccinated commercial and backyard poultry in Egypt' *Poultry Science* 2010; 89: 1609–1613.
243. Mackenzie D. 'Bird flu vaccination could lead to new strains' *New Scientist* March 24, 2004.
244. Lee C-W, Senne DA, Suarez DL. 'Effect of vaccine use in the evolution of Mexican lineage H5N2 avian influenza virus' *Journal of Virology* 2004; 78: 8372–8381.
245. WHO, 'H5N1 avian influenza: Timeline', 2011.
246. Martinot et al., 2007.
247. Tomley FM, Shirley MW. 'Livestock infectious diseases and zoonoses' *Philosophical Transactions of the Royal Society B Biological Sciences* 2009; 364: 2637–2642.
248. Taubenberger and Morens, 2006.

249. Normille D. 'Flu virus research yields results but no magic bullet for pandemic' *Science* 2008; 319: 1178–1179.

5 Animal Agriculture: Our Health and Our Environment

1. Food and Agriculture Organization (FAO). 'Livestock a major threat to environment' November 29, 2006. www.fao.org, date accessed December 1, 2010.
2. Schmidt CW. 'Lessons from the flood: Will Floyd change livestock farming?' *Environmental Health Perspectives* 2000; 108: A74–A77.
3. Ibid.; Robert Wood Johnson Foundation. 'Flooding from hurricane Floyd produced negative health consequences from farm practices' August 2007. www.rwjf.org, date accessed December 5, 2010.
4. Robert Wood Johnson Foundation, 2007.
5. Schmidt, 2000; Environmental Defense Fund. 'Cleaning up hog waste in North Carolina. Working to show hog farming can be clean and profitable, updated September, 2007'. www.edf.org, date accessed December 6, 2010.
6. Steinfeld H, Gerber P, Wassenaar T et al. *Livestock's Long Shadow. Environmental Issues and Options* (Rome: Food and Agriculture Organization of the United Nations) 2006.
7. Costanza JK, Marcinko SE, Goewert AE, Mitchell CE. 'Potential geographic distribution of atmospheric nitrogen deposition from intensive livestock production in North Carolina, USA' *Science of the Total Environment* 2008; 398: 76–86; Carpenter SR, Caraco NF, Correll DL et al. 'Nonpoint pollution of surface waters with phosphorus and nitrogen' *Ecological Applications* 1998; 8: 559–568.
8. National Oceanic and Atmospheric Administration. 'Investigating the ocean algae blooms'. www.science-house.org, date accessed December 7, 2010.
9. Blazer V, Phillips S, Pendleton E. 'Fish health, fungal infections, and *Pfiesteria*: The role of the U.S. Geological Survey' *US Geological Survey Fact Sheet 114–98*. pubs.usgs.gov/fs/1998/114/, date accessed December 2, 2010.
10. Collier DN, Burke WA. '*Pfiesteria* complex organisms and human illness' *Southern Medical Journal* 2002; 95: 720–726.
11. Copeland C. 'Animal waste and water quality: EPA regulation of Concentrated Animal Feeding Operations (CAFOs)' *Environmental Protection Agency*, updated November 17, 2008. www.nationalaglawcenter.org, date accessed December 7, 2010; Majumdar G. 'The blue baby syndrome: Nitrate poisoning in infants' *Humanities, Social Sciences and Law* 2003; 8: 20–30.
12. FAO, 2006.
13. US Environmental Protection Agency. 'National pollutant discharge elimination system permit regulations and effluent limitation guideline and standards for concentrated animal feeding operations (CAFOs); final rule' *Federal Register* February 12, 2003; 68 (29): 7237
14. Copeland, 2008; US Environmental Protection Agency, 2003.
15. Copeland, 2008.
16. Ibid.; Constanza et al., 2008.

17. Battaye R, Overcash C, Fudge S. *Development and Selection of Ammonia Emission Factors* (Washington, DC: US Environmental Protection Agency) 1994. Nepis.epa.gov, date accessed December 7, 2010.
18. Thu K. (ed.) *Understanding the Impacts of Large-Scale Swine Production: Proceedings from an Interdisciplinary Scientific Workshop* (Des Moines, IA) June 29–30, 1995; Steinfeld et al., 2006; American Public Health Association (APHA). 'Precautionary Moratorium on New Concentrated Animal Feed Operations Policy number 20037, 1/18/2003'. www.apha.org, date accessed October 11, 2010.
19. Wing S, Wolf S. 'Intensive livestock operations, health, and quality of life among Eastern North Carolina residents' *Environmental Health Perspectives* 2000; 108: 233–238.
20. Sigurdarson St, Kline JN. 'School proximity to concentrated animal feeding operations and prevalence of asthma in students' *CHEST* 2006; 129: 1486–1491.
21. Mirabelli MC, Wing S, Marshall S, Wilcosky T. 'Asthma symptoms among adolescents who attend public schools that are located near confined swine feeding operations' *Pediatrics* 2006; 118: e66–e75.
22. Schiffman SS, Miller EAS, Suggs MS, Graham BG. 'The effect of environmental odors emanating from commercial swine operations on the mood of nearby residents' *Brain Research Bulletin* 1995; 37: 369–375.
23. Schiffman SS. 'Livestock odors: Implications for human health and well-being' *Journal of Animal Science* 1998; 76: 1343–1355.
24. Radon K, Peters A, Praml G et al. 'Livestock odours and quality of life of neighboring residents' *Annals of Agricultural and Environmental Medicine* 2004; 11: 59–62.
25. Horton RA, Wing S, Marshall SW, Brownley KA. 'Malador as a trigger of stress and negative mood in neighbors of industrial hog operations' *American Journal of Public Health* 2009; 99: S610–S615; Radon K, Schulze A, Ehrenstein V et al. 'Environmental exposure to confined animal feeding operations and respiratory health of neighboring residents' *Epidemiology* 2007; 18; 300–308; Villeneuve PJ, Ali A, Challacombe L, Herbert S. 'Intensive hog farming operations and self-reported health among nearby rural residents in Ottawa, Canada' *BMC Public Health* 2009; 9: 330; Wing S, Horton RA, Marshall SW et al. 'Air pollution and odor in communities near industrial swine operations' *Environmental Health Perspectives* 2008; 116: 1362–1368; Donham KJ, Wing S, Osterberg D et al. 'Community health and socioeconomic issues surrounding concentrated animal feeding operations' *Environmental Health Perspectives* 2007; 115: 317–320; Heederik D, Sigsgaard T, Thorne PS et al. 'Health effects of airborne exposures from concentrated animal feeding operations' *Environmental Health Perspectives* 2007; 115: 298–302.
26. Wing S, Cole D, Grant G. 'Environmental injustice in North Carolina's hog industry' *Environmental Health Perspectives* 2000; 108: 225–231; Wilson SM, Howell F, Wing S, Sobsey M. 'Environmental injustice and the Mississippi hog industry' *Environmental Health Perspectives* 2002; 110: 195–201.
27. Horton et al., 2009.
28. Steinfield, 2006.
29. Ibid.

30. Ibid.
31. Aneja VP, Schlesinger WH, Erisman JW. 'Effects of agriculture upon the air quality and climate: Research, policy, and regulations' *Environmental Science and Technology* 2009; 43: 4234–4240.
32. New York Times. 'Science watch: Acid rain enhancer' *New York Times* September 25, 1990.
33. Goodland R, Anhang J. 'Livestock and climate change: What if the key actors in climate change are cows, pigs, and chickens?' *WorldWatch* 2009 November/December. www.worldwatch.org, date accessed November 9, 2010; Mackay F. 'Solving the problem of methane and meat; scientists explore nature and technology to cut bovine gas emissions' *International Herald Times* November 17, 2009.
34. Fanelli D. 'Meat is murder on the environment' *New Scientist* 2007; 195: 15.
35. Steinfeld et al., 2006.
36. Ibid.
37. Ibid.
38. Pimentel D, Pimentel M. 'Sustainability of meat-based and plant-based diets and the environment' *American Journal of Clinical Nutrition* 2003; 78: 660S–663S.
39. Fiala N. 'The greenhouse hamburger' *Scientific American* 2009; 300: 72–75.
40. Costello A, Abbas M, Allen A et al. 'Managing the health effects of climate change; *Lancet* and University College London Institute for Global Health Commission' *Lancet* 2009; 373: 1693-1773.
41. National Oceanic and Atmospheric Administration (NOAA). 'National Climatic Data Center. Global warming. Frequently asked questions'. www.ndcd.noaa.gov, date accessed December 5, 2010.
42. Bernstein L, Bosch P, Osvaldo C et al. 'Climate change 2007: Synthesis report. Summary for policy makers. An assessment of the Intergovernmental Panel on Climate Change' *Intergovernmental Panel on Climate Change.* www.eldis.org, date accessed December 6, 2010.
43. Ibid.
44. Ibid.; Luber G, Prudent N. 'Climate change and human health' *Transactions of the American Clinical and Climatological Association* 2009; 120: 113–117.
45. Bernstein et al., 2007; Costello et al., 2009; Kjellstrom T, Butler AJ, Lucas RM, Bonita R. 'Public health impact of global heating due to climate change: Potential effects on chronic, non-communicable diseases' *International Journal of Public Health* 2010; 55: 97–103; Luber and Prudent, 2009; Rosenthal JP, Jessup CM. 'Global climate change and health: Developing a research agenda for NIH' *Transactions of the American Clinical and Climatological Association* 2009; 120: 129–139.
46. Bernstein et al., 2007.
47. Steinfeld et al., 2006.
48. Costello et al., 2009.
49. Rosenthal and Jessup, 2009; Kjellstrom et al., 2009; Sarfaty M, Abouzaid S. 'The physician's response to climate change' *Family Medicine* 2009; 41: 358–363.
50. Costello et al., 2009.
51. Costello A. 'The Lancet-UCL Commission: Health effects of climate change' *Lancet* 2008; 371: 1145–1147.

52. Costello et al., 2009.
53. Frumkin H, Hess J, Luber G, Malilay J, McGeehin M. 'Climate change: The public health response' *American Journal of Public Health* 2008; 98:435–445; Shea KM. 'American Academy of Pediatrics Committee on Environmental Health: Global climate change and children's health' *Pediatrics* 2007; 120: e1359–e1367; Bernstein et al., 2007; Frumkin H, McMichael AJ, Hess JJ. 'Climate change and the health of the public' *American Journal of Preventive Medicine* 2008; 35: 401–402.
54. Ezzati M, Lopez AD, Rodgers A, Vander Hoorn S, Murray CJL. 'Comparative Risk Assessment Collaborating Group: Selected major risk factors and global and regional burden of disease' *Lancet* 2002; 360: 1347–1360.
55. Conisbee M, Simms A. 'Environmental refugees: The case for recognition' *New Economics Foundation* September 30, 2003. www.neweconomics.org/publications/environmental-refugees, date accessed January 1, 2011.
56. Food and Agriculture Organization (FAO). 'Livestock a major threat to the environment' November 29, 2006. http://www.fao.org/newsroom/en/news/2006/1000448/index.html, date accessed October 10, 2010.
57. APHA, 2003.
58. Weiss R. 'Report target costs of factory farming' *Washington Post* August 30, 2008.
59. Faergeman O. 'Climate change and preventive medicine' *European Journal of Cardiovascular Prevention and Rehabilitation* 2007; 14: 726–729.
60. Marlow HJ, Hayes WK, Soret S et al. 'Diet and the environment; Does what you eat matter?' *American Journal of Clinical Nutrition* 2009; 89: 1699S–1703S.
61. Akhtar A, Greger M, Ferdowsian H, Frank E. 'Health professionals' roles in animal agriculture, climate change, and human health' *American Journal of Preventive Medicine* 2009; 36: 182–187.
62. McMichael AJ, Powles JW, Butler CD, Uauy R. 'Food, livestock production, energy, climate change and health' *Lancet* 2007 370: 1253–1263.
63. Carlsson-Kanyama A, Gonzalez AD. 'Potential contributions of food consumption patterns to climate change' *American Journal of Clinical Nutrition* 2009; 89: 1704S–1709S.
64. Marlow et al., 2009.
65. Powles J. 'Commentary: Why diets need to change to avert harm from global warming' *International Journal of Epidemiology* 2009; 38: 1141–1142; Jacobs DR, Haddah EH, Lanou AJ, Messina MJ. 'Food, plant food, and vegetarian diets in the US dietary guidelines: conclusions of an expert panel' *American Journal of Clinical Nutrition* 2009; 89 (Suppl.): 1549S–1552S; Carlsson-Kanyama and Gonzalez, 2009; McMichael et al., 2007; Marlow et al., 2009; Neff RA, Chan IL, Smith KC. 'Yesterday's dinner, tomorrow's weather, today's news? US newspaper coverage of food system contributions to climate change' *Public Health Nutrition* 2008; 12: 1006–1014; Pimentel D, Pimentel M. 'Sustainability of meat-based and plant-based diets and the environment' *American Journal of Clinical Nutrition* 2003; 78: 660S–663S; Prigg M, Goodchild S. 'Eat less meat to stop climate change' *Evening Standard* (London) November 25, 2009.
66. Costello et al., 2009.
67. Steinfeld et al., 2006.

68. Popkin BM. 'Reducing meat consumption has multiple benefits for the world's health' *Archives of Internal Medicine* 2009; 169: 543–545.
69. Friel S, Dangour AD, Garnett T et al. 'Public health benefits of strategies to reduce greenhouse-gas emissions: Food and agriculture' *Lancet* 2009; 374: 2016–2025.
70. Yach D, Hawkes C, Gould CL, Hofman KJ. 'The global burden of chronic diseases. Overcoming impediments to prevention and control' *Journal of the American Medical Association* 2004; 291: 2616–2622.
71. World Health Organization (WHO). *The Global Burden of Disease: 2004 Update* (Geneva: World Health Organization) 2008.
72. Yach et al., 2004.
73. WHO 'The global burden of disease: *2004 Update*', 2008; Yach et al., 2004.
74. Yach et al., 2004.
75. Popkin BM, Du S. 'Dynamics of the nutrition transition towards the animal foods sector in China and its implications: A worried perspective' *Journal of Nutrition* 2003; 133: 3898S–3906S.
76. World Health Organization (WHO). 'Obesity and overweight Factsheet'. www.who.int, date accessed March 13, 2008; Centers for Disease Control and Prevention (CDC). 'Prevalence of overweight and obesity among adults: United States, 2003–2004' *National Center for Health Statistics Factsheet*. www.cdc.gov, date accessed April 1, 2008.
77. WHO, 'Obesity and overweight'.
78. Mokdad AH, Ford ES, Bowman BA et al. 'Prevalence of obesity, diabetes, and obesity-related health risk factors, 2001' *Journal of the American Medical Association* 2003; 289: 76–79.
79. Doheny K. 'U.S. obesity rate may hit 42% by 2050' *HealthDay Reporter* November 5, 2010.
80. Popkin and Du, 2003; Walker P, Rhubart-Berg P, McKenzie S, Kelling K, Lawrence RS. 'Public health implications of meat production and consumption' *Public Health Nutrition* 2005; 8: 348–356.
81. Xu W-H, Dai Q, Xiang Y-B et al. 'Nutritional factors in relation to endometrial cancer: A report from a population-based case-control study in Shanghai, China' *International Journal of Cancer* 2007; 120: 1776–1781.
82. Mitrou PN, Albanes D, Weinstein SJ et al. 'A prospective study of dietary calcium, dairy products and prostate cancer risk (Finland)' *International Journal of Cancer* 2007; 120: 2466–2473; Chan JM, Stampfer MJ, Ma J et al. 'Dairy products, calcium, and prostate cancer risk in the Physician's Health Study' *American Journal of Clinical Nutrition* 2001; 74: 549–554; Kurahashi N, Inoue M, Iwasaki M, Sasazuki S, Tsugane AS. 'For the Japan public health center-based prospective study group: Dairy product, saturated fatty acid, and calcium intake and prostate cancer in a prospective cohort of Japanese men' *Cancer, Epidemiology, Biomarkers and Prevention* 2008; 17: 930–937.
83. Allen NE, Key TJ, Appleby PN et al. 'Animal foods, protein, calcium and prostate cancer risk: The European Perspective Investigation into Cancer and Nutrition' *British Journal of Cancer* 2008; 6: 1574–1581.
84. Taylor EF, Burley VJ, Greenwood DC, Cade JE. 'Meat consumption and risk of breast cancer in the UK Women's Cohort Study' *British Journal of Cancer* 2007; 96: 1139–1146; Cho E, Chen WY, Hunter DJ et al. 'Red meat intake and risk of breast cancer among premenopausal women' *Archives*

of Internal Medicine 2006; 166: 2253–2259; Boyd NF, Stone J, Vogt KN et al. 'Dietary fat and breast cancer risk revisited: A meta-analysis of the published literature' *British Journal of Cancer* 2003; 89: 1672–1685; Linos E, Willett WC, Cho E, Colditz G, Frazier LA. 'Red meat consumption during adolescence among premenopausal women and risk of breast cancer' *Cancer Epidemiology, Biomarkers and Prevention* 2008; 17: 2146–2151.

85. Armstrong B, Doll R. 'Environmental factors and cancer incidence and mortality in different countries, with special reference to dietary practices' *International Journal of Cancer* 1975; 15: 617–631.

86. Kuriki K, Tajima K. 'The increasing incidence of colorectal cancer and the preventive strategy in Japan' *Asian Pacific Journal of Cancer Prevention* 2006; 7: 495–501; Kuriki K, Tokudome S, Tajima K. 'Association between type II diabetes and colon cancer among Japanese with reference to change in food intake' *Asian Pacific Journal of Cancer Prevention* 2004; 5: 28–35.

87. Chao A, Thun MJ, Connell CJ et al. 'Meat consumption and risk of colorectal cancer' *Journal of the American Medical Association* 2005; 293: 172–182; Cross AJ, Leitzmann MF, Gail MH et al. 'A prospective study of red and processed meat intake in relation to cancer risk' *PLoS Medicine* 2007; 4: e325; World Cancer Research Fund/American Institute for Cancer Research (WCF/AIR). *Food, Nutrition, Physical Activity and the Prevention of Cancer: A Global Perspective* (Washington, DC: AICR) 2007; Lee SI, Moon HY, Kwak JM et al. 'Relationship between meat and cereal consumption and colorectal cancer in Korea and Japan' *Journal of Gastroenterology and Hepatology* 2008; 23: 138–140.

88. WCF/AICR, 2007.

89. Popkin and Du, 2003; Sluijs I, Beulens JWJ, Van Der ADL, et al. 'Dietary intake of total, animal; and vegetable protein and risk of type 2 diabetes in the European Prospective Investigation into Cancer and Nutrition (EPIC)-NL study' *Diabetes Care 2010*; 33: 43–48; Lin Y, Bolca S, Vandevijvere S et al. 'Plant and animal protein intake and its association with overweight and obesity among the Belgian population' *British Journal of Nutrition* December 9, 2010: 1–11; Halkjaer J, Olsen A, Jakobsen MU et al. 'Intake of total, animal and plant protein and subsequent changes in weight or waist circumference in European men and women: The Diogenes project' *International Journal of Obesity (London)* December 7, 2010 (epub ahead of print).

90. Barnard ND, Nicholson A, Howard JL. 'The medical costs attributable to meat consumption' *Preventive Medicine* 1995; 24: 646–655.

91. Preidt R. 'Estimated cost of obesity is $300 billion per year' *HealthDay News* January 11, 2011.

92. Reinberg S. 'U.S. heart disease costs expected to soar' *HealthDay News* January 24, 2011.

93. Berkow SE, Barnard N. 'Vegetarian diets and weight status' *Nutrition Reviews* 2006; 64: 175–188; Melby CL, Hyner GC, Zoog B. 'Blood pressure in vegetarians and non-vegetarians: A cross-sectional analysis' *Nutrition Research* 1985; 5:1077–1082; Sabaté J, Wien M. 'Vegetarian diets and childhood obesity prevention' *American Journal of Clinical Nutrition* 2010 ; 91 (5): 1525S–1529S; American Dietetic Association. 'Position of the American Dietetic

Association: Vegetarian diets' *Journal of the American Dietetic Association* 2009; 109: 1266–1282.

94. Ornish D, Scherwitz LW, Billings JH et al. 'Intensive lifestyle changes for reversal of coronary heart disease' *Journal of the American Medical Association* 1998; 280: 2001–07; Kwok TK, Woo J, Ho S, Sham A. 'Vegetarianism and ischemic heart disease in older Chinese women' *Journal of American College of Nutrition* 2000; 19: 622–627; Key TJ, Fraser GE, Thorogood M et al. 'Mortality in vegetarians and non-vegetarians: A collaborative analysis of 8300 deaths among 76,000 men and women in five prospective studies' *Public Health Nutrition* 1998; 1: 33–41; Barnard ND, Cohen J, Jenkins DJ et al. 'A low-fat vegan diet improves glycemic control and cardiovascular risk factors in a randomized clinical trial in individuals with type 2 diabetes' *Diabetes Care* 2006; 29: 1777–1783; Saxe GA, Major JM, Nguyen JY et al. 'Potential attenuation of disease progression in recurrent prostate cancer with plant-based diets and stress reduction' *Integrative Cancer Therapies* 2006; 5: 206–213; Ornish D, Weidner G, Fair WR et al. 'Intensive lifestyle changes may affect the progression of prostate cancer' *Journal of Urology* 2005; 174: 1065–1069; Cui X, Dai Q, Tseng M, Shu XO, Gao YT, Zheng W. 'Dietary patterns and breast cancer risk in the Shanghai breast cancer study' *Cancer Epidemiology, Biomarkers and Prevention* 2007; 16: 1443–1448.
95. American Dietetic Association, 2009.
96. Katz DL. 'Plant foods in the American diet? As we sow ...' *Medscape Journal of Medicine* 2009; 11: 25.
97. World Health Organization (WHO). *Global Health Risks. Mortality and Burden of Disease Attributable to Selected Major Risks* (Geneva: World Health Organization) 2009.
98. Mokdad AH, Marks JS, Stroup DF, Gerberding JL. 'Actual causes of death in the United States, 2000' *Journal of the American Medical Association* 2004; 291: 1238–1245.
99. Akhtar et al., 2009.
100. Ibid.
101. Zatonski WA, McMichael AJ, Powles JW. 'Ecological study of reasons for sharp decline in mortality from ischemic heart disease in Poland since 1991' *British Medical Journal* 1998; 316: 1047–1051.
102. Prigg and Goodchild, 2009.
103. Akhtar et al., 2009.
104. American Public Health Association (APHA). 'Climate change: Our health in the balance. Healthy climate pledge' 2008. www.ecdh.org, date accessed January 6, 2010; Johns Hopkins Bloomberg School of Public Health, Center For a Livable Future Programs. 'Eating for the future. Meatless Monday Campaign, Inc'. www.jhsphu.edu, date accessed December 6, 2010.
105. Harvey J. 'Menu of change: Healthy foods in health care. A 2008 survey of healthy food in health care pledge hospitals' *Health Care Without Harm* 2008. www.noharm.org, date accessed January 6, 2009.
106. Balcombe J. 'From skinny bitch to Bill Clinton: The rise of veganism' *Psychology Today* December 14, 2010. www.psychologytoday.com, date accessed December 18, 2010; Seth S. 'Rise of power vegan' *Franchise India* November 11, 2010; Rogers T. 'Forget a vegan, he's a "hegan"'.

Salon.com March 24, 2010; Betty Crocker. 'Top ten trends' *Betty Crocker.* redhot.bettycrocker.com, date accessed December 10, 2010.
107. Kondro W. 'Canada needs a paradigm shift in public health nutrition' *Canadian Medical Association Journal* 2008; 179: 1259–1261.

6 The Costs of Animal Experiments

1. Dorsey ER, de Roulet J, Thompson JP et al. 'Funding of US biomedical research, 2003–2008' *Journal of the American Medical Association* 2010; 303: 137–143; Roberts J. 'Spending on medical research soars. U.S. now spends $95 billion a year on research, study finds' *CBS News* September 20, 2005. www.cbsnews.com/stories/2005/09/20/health/main861059. shtml, date accessed December 10, 2010; Moses H. 'Researchers, funding, and priorities: The razor's edge' *Journal of the American Medical Association* 2009; 302: 1001–1002; Philipson L. 'Medical research activities, funding, and creativity in Europe: Comparison with research in the United States' *Journal of the American Medical Association* 2005; 294: 1394–1398.
2. Reuters. 'U.S. medical research funding falls: Analysis' January 12, 2010. www.reuters.com, date accessed January 13, 2010; Sheridan K. 'US scientists sound alarm over animal research' January 6, 2011. Physorg.com, date accessed January 11, 2011; Dorsey et al., 2010.
3. Dorsey et al., 2010.
4. MacAskill E. 'U.S. tumbles down in world ratings list for life expectancy' *Guardian* August 13, 2007.
5. Hampton T. 'Targeted cancer therapies lagging. Better trial design could boost success rate' *Journal of the American Medical Association* 2006; 296: 1951–1952.
6. Ibid.
7. Gitig D. 'Increasing percentage of new drugs are failing phase II and III trials' *Highlight Health* June 1, 2011. www.highlighthealth.com, date accessed June 10, 2011.
8. Begley S, Carmichael M. 'Desperately seeking cures' *Newsweek* May 14, 2010.
9. Persson CGA, Erjefalt JS, Uller L, Andersson M, Greiff L. 'Unbalanced research' *Trends in Pharmacological Sciences* 2001; 22: 538–541; Horrobin DF. 'Innovation in the pharmaceutical industry' *Journal of the Royal Society of Medicine* 2000; 93: 341–345; Moses H, Dorsey ER, Matheson DHM, Thier SO. 'Financial anatomy of biomedical research' *Journal of the American Medical Association* 2005; 294: 1333–1342.
10. Langreth R. 'Many cancer drugs, hardly any breakthroughs' *Forbes Science Business* June 6, 2010. www.forbes.com, date accessed January 3, 2011.
11. Dorsey et al., 2010.
12. Rosenberg RN. 'Translating biomedical research to the bedside: A national crisis and call to action' *Journal of the American Medical Association* 2003; 289: 1305–1306.
13. Ibid.
14. Ioannidis JPA. 'Evolution and translation of research findings: From bench to where?' *PLoS Clinical Trials* 2006; 1: e36; O'Connell D, Roblin

D. 'Translational research in the pharmaceutical industry: From bench to bedside' *Drug Discovery Today* 2006; 11: 833–838.

15. Johnston CS. 'Translation: Case study in failure' *Annals of Neurology* 2006; 59: 447–448.
16. Hurko O, Ryan JL. 'Translational research in central nervous system drug discovery' *NeuroRx* 2005; 2: 671–682.
17. Johnston, 2006; Wall RJ, Shani M. 'Are animal models as good as we think?' *Theriogenology* 2008; 69: 2–9.
18. *Data from Experimental Stroke (CAMARADES).* www.camarades.info, date accessed June 10, 2011.
19. Ibid.
20. Macleod MR, O'Collins T, Howells DW, Donnan GA. 'Pooling of animal experimental data reveals influence of study design and publication bias' *Stroke* 2004; 35: 1203–1208.
21. Wall and Shani, 2008.
22. Ibid.
23. Palfreyman MG, Charles V, Blander J. 'The importance of using human-based models in gene and drug discovery' *Drug Discovery World* Fall 2002: 33–40.
24. Sinha G. 'Cell-based tests tackle predicting safety of antibody drugs' *Nature Medicine* 2006; 12: 485; Allini M, Eisenstein SM, Ito K et al. 'Are animal models useful for studying human disc disorders/degeneration? *European Spine Journal* 2008; 17: 2–19; Horrobin DF. 'Modern biomedical research: An internally self-consistent universe with little contact with medical reality?' *Nature Reviews Drug Discovery* 2003; 2: 151–154; New Scientist. 'Brainstorming' *New Scientist* October 16, 2004: 2469. www.newscientist.com, date accessed January 14, 2011; Wall and Shani, 2008.
25. Matthews RJ. 'Medical progress depends on animal models—Doesn't it?' *Journal of the Royal Society of Medicine* 2008; 101: 95–98.
26. Archibald K. 'No need for monkeys' *New Scientist* 2006; 191: 26; Neyt JG, Buckwalter JA, Carroll NC. 'Use of animal models in musculoskeletal research' *Iowa Orthopedic Journal* 1998; 18: 118–123; Sams-Dodd F. 'Strategies to optimize the validity of disease models in the drug discovery process' *Drug Discovery Today* 2006; 11: 355–363; Allen A. 'Of mice and men: The problems with animal testing' *Slate* June 1, 2006.
27. Pound P, Ebrahim S. 'Supportive evidence is lacking on animal studies' *British Medical Journal* 2002; 325: 1038.
28. Nuffield Council on Bioethics. 'The ethics of research involving animals' May 25, 2005. www.nuffieldbioethics.org, date accessed December 10, 2010.
29. Pound P, Ebrahim S, Sandercock P et al. 'Where is the evidence that animal research benefits humans?' *British Medical Journal* 2004; 328: 514–517; Matthews, 2008.
30. Varga OE, Hansen AK, Sandoe P, Olsson IA. 'Validating animal models for preclinical research: A scientific and ethical discussion' *Alternative to Laboratory Animals* 2010; 38: 245–248.
31. Neyt, 1998.
32. Allen, 2006.
33. New Scientist, 2004.

34. Akhtar A, Pippin JJ, Sandusky CB. 'Animal studies in spinal cord injury: A systematic review of methylprednisolone' *Alternatives to Laboratory Animals* 2009; 37: 43–62; Akhtar A, Pippin JJ, Sandusky CB. 'Animal models in spinal cord injury: A review' *Reviews in the Neurosciences* 2008; 19: 47–60; Kwon BJ, Hillyer J, Tetzlaff W. 'Translational research in spinal cord injury: A survey of opinion from the SCI community' *Journal of Neurotrauma* 2010; 27: 21–33.
35. Akhtar et al., 2009.
36. Akhtar et al., 2008.
37. Baldwin A. Bekoff M. 'Too stressed to work' *New Scientist* 2007; 194: 24.
38. Yevgenia K, Gross CG, Kopil C et al. 'Experience induces structural and biochemical changes in the adult primate brain' *Proceedings of the National Academy of Sciences* 2005; 102: 17478–17482.
39. O' Neil BJ, Kline JA, Burkhart K, Younger J. 'Research fundamentals: V. The use of laboratory animal models in research' *Academic Emergency Medicine* 1999; 6: 75–82; Balcombe JP, Barnard ND, Sandusky C. 'Laboratory routines cause animal stress' *Contemporary Topics in Laboratory Animal Science* 2004; 43: 42–51.
40. Akhtar et al., 2008; Balcome et al., 2004; Wilson LM, Baldwin AL. 'Environmental stress causes mast cell degranulation, endothelial and epithelial changes, and edema in rat intestinal mucosa' *Microcirculation* 1999; 6: 189–198; Baldwin and Bekoff, 2007.
41. Baldwin and Bekoff, 2007.
42. Kriegsfeld LJ, Eiasson MJ, Demas GE et al. 'Nocturnal motor coordination deficits in neuronal nitric oxide synthase knock-out mice' *Neuroscience* 1999; 89: 311–315.
43. O'Neil et al., 1999.
44. Chesler EJ, Wilson SG, Lariviere WR, Rodriguez-Zas SL, Mogil JS. 'Identification and ranking of genetic and laboratory environment factors influencing a behavioral trait, thermal nociception, via computational analysis of a large data archive' *Neuroscience & Biobehavioral Reviews* 2002; 26: 907–923.
45. Gawrylewski A. 'The trouble with animal models. Why did human trials fail?' *The Scientist* 2007; 21: 44.
46. Ibid.
47. Akhtar et al., 2009; Lonjon N, Prieto M, Haton H et al. 'Minimum information about animal experiments: Supplier is also important' *Journal of Neuroscience Research* 2009; 87: 403–407; O'Neil et al., 1999; MacLeod et al., 2004.
48. Balcombe et al., 2004.
49. Crabbe JC, Wahlsten D, Dudek BC. 'Genetics of mouse behavior: Interactions with laboratory environment' *Science* 1999; 284: 1670–1672.
50. Sandercock P, Roberts I. 'Systematic reviews of animal experiments' *Lancet* 2002; 360: 586; Macleod M, Sandercock P. 'Can systematic reviews help animal experiments work?' *RDS News* Winter 2005; Roberts I, Kwan I, Evans P, Haig S. 'Does animal experimentation inform human health care? Observations from a systematic review of international animal experiments on fluid resuscitation' *British Medical Journal* 2002; 324: 474–476; *CAMARADES*, date accessed December 30, 2010; Gawrylewski, 2007.
51. Akhtar et al., 2009.

52. Gawrylewski, 2007.
53. Ibid.
54. Ibid.
55. Curry SH. 'Why have so many drugs with stellar results in laboratory stroke models failed in clinical trials? A theory based on allometric relationships' *Annals of the New York Academy of Sciences* 2003; 993: 69–74; Dirnagl U. 'Bench to bedside: The quest for quality in experimental stroke research' *Journal of Cerebral Blood Flow & Metabolism* 2006; 26: 1465–1478.
56. van der Worp HB, Howells DW, Sena ES et al. 'Can animal models of disease reliably inform human studies?' *PLoS Medicine* 2010; 7 (3).
57. Dirnagl, 2006; Macleod et al., 2004; Sena E, van der Worp B, Howells D, Macleod M. 'How can we improve the pre-clinical development of drugs for stroke?' *Trends in Neurosciences* 2007; 30: 433–439.
58. Wiebers DO, Adams HP, Whisnant JP. 'Animal models of stroke: Are they relevant to human disease?' *Stroke* 1990; 21: 1–3.
59. Ibid.
60. Hurko and Ryan, 2005.
61. Fallon L. 'Drug discovery for neurodegeneration—Inaugural Alzheimer's drug discovery foundation meeting' *IDrugs* 2007; 10: 233–236.
62. McGowan E, Eriksen J, Hutton M. 'A decade of modeling Alzheimer's disease in transgenic mice' *TRENDS in Genetics* 2006; 22: 281–289; Balducci C, Forloni G. 'APP transgenic mice: Their use and limitations' *Neuromolecular Medicine* December 9, 2010 (epub ahead of print); Jay GW, Memattos RB, Weinstein EJ et al. 'Animal models for neural diseases' *Toxicologic Pathology* November 30, 2010 (epub ahead of print); Götz J, Ittner LM. 'Animal models of Alzheimer's disease and frontotemporal dementia' *Nature* 2008; 9: 532–544; Fallon, 2007.
63. Medical News Today. 'Effectiveness of mouse breeds that mimic Alzheimer's disease symptoms questioned' *Medical News Today* August 17, 2007. www.medicalnewstoday.org, date accessed January 1, 2011.
64. Ibid.
65. Fallon, 2007.
66. Sams-Dodd, 2006.
67. Horrrobin, 2003.
68. Cabrera O, Berman DM, Kenyon NS et al. 'The unique cytoarchitecture of human pancreatic islets has implications for islet cell function' *Proceedings of the National Academy of Sciences* 2006; 103: 2334–2339; Kakulas BA. 'The applied neuropathology of human spinal cord injury' *Spinal Cord* 1999; 37: 79–88; Akhtar et al., 2009.
69. New Scientist, 2004.
70. Akhtar et al., 2008; Lonjon et al., 2009.
71. Akhtar et al., 2008.
72. Lonjon et al., 2009.
73. Mogil JS, Wilson SG, Bon K et al. 'Heritability of nociception I: Responses of 11 inbred mouse strains on 12 measures of nociception' *Pain* 1999; 80: 67–82.
74. Coleman RA. 'Asterand PhaseZERO—A solution for drug development. Next generation pharmaceutical' July 6, 2010. www.ngpharma.com, date accessed January 3, 2011; Balducci and Forloni, 2010.

75. Terszowski G, Muller SM, Bleul CC et al. 'Evidence for a functional second thymus in mice' *Science* 2006; 312: 284–287.
76. Diabetes Research Institute. 'Researchers find striking differences between human and animal insulin-producing islet cells' February 2006. www.diabetesresearch.org, date accessed December 12, 2010; Cabrera et al., 2006.
77. Palfreyman et al., 2002.
78. Laboratory News. 'New guidlines published as mouse models thrown into question' June, 2008. www.labnews.co.uk, date accessed December 27, 2010; Liao B-Y, Zhang J. 'Null mutations in humans and mouse orthologs frequently result in different phenotypes' *Proceedings of the National Academy of Sciences* 2008; 105: 6987–6992.
79. Liao and Zhang, 2008.
80. Laboratory News, 2008.
81. Odom DT, Dowell RD, Jacobsen ES et al. 'Tissue-specific transcriptional regulation has diverged significantly between human and mouse' *Nature Genetics* 2007; 39: 730–732.
82. Horrobin, 2003.
83. Ibid.
84. Ibid.
85. Benatar M. 'Lost in translation: Treatment trials in the SOD1 mouse and in human ALS' *Neurobiology of Disease* 2007; 26: 1–13; Gawryleski, 2007.
86. Coleman, 2010.
87. Linder S, Shoshan MC. 'Is translational research compatible with preclinical publication strategies?' *Radiation Oncology* 2006; 1: 4.
88. Leslie M. 'Immunology uncaged' *Science* 2010; 327: 1573.
89. Ledford H. 'Flaws found in mouse model of diabetes' *Nature* May 28, 2009; 523.
90. Guttman-Yassky E, Krueger JG. 'Psoriasis: Evolution of pathogenic concepts and new therapies through phases of translational research' *British Journal of Dermatology* 2007; 157: 1103–1115; New Scientist, 2004; Persson CGA, Erjefalt JS, Korsgren M, Sundler F. 'The mouse trap' *Trends in Pharmacological Sciences* 1997; 18: 465–467.
91. Allen, 2006.
92. Ibid.
93. Ibid.
94. Ibid.
95. Attarwala H. 'TGN1412: From discovery to disaster' *Journal of Young Pharmacists* July–September 2010; 2 (3): 332–336; Hanke T. 'Lessons from TGN1412' *Lancet* 2006; 368: 1569–1570.
96. Attarwala, 2010.
97. Fraga MF, Ballestar E, Paz MF et al. 'Epigenetic differences arise during the lifetime of monozygotic twins' *Proceedings of the National Academy of Sciences* 2005; 102: 10604–10609.
98. Ahmed J., Günther S, Möller F, Preissner R. 'A structural genomics approach to the regulation of apoptosis: Chimp vs. human' *Genome Informatics* 2007; 18: 22–34; Palfreyman et al., 2002; Khaitovich P, Muetzel B, She X et al. 'Regional patterns of gene expression in human and chimpanzee brains' *Genome Research* 2004; 14: 1462–1473; Normille D. 'Gene expression differs in human and chimp brains' *Science* 2001; 292: 44–45.

99. Ahmed et al., 2007.
100. Lane E, Dunnett S. 'Animal models of Parkinson's disease and L-dopa induced dyskinesia: How close are we to the clinic?' *Psychopharmacology* 2008; 199: 303–312.
101. Ibid.
102. Hogan RJ. 'Are nonhuman primates good models for SARS?' *PLoS Medicine* 2006; 3: 1656–1657.
103. Ibid.
104. Bettauer RH. 'Chimpanzees in hepatitis c virus research: 1998–2007' *Journal of Medical Primatology* 2010; 39: 9–23; Bailey J. 'An assessment of the use of chimpanzees in hepatitis c research past, present and future: 1. Validity of the chimpanzee model' *ATLA* 2010; 38: 387–418.
105. Tonks A. 'The quest for the AIDs vaccine' *British Medical Journal* 2007; 334: 1346–1348.
106. Archibald, 2006; Bailey J. 'An assessment of the role of chimpanzee in AIDS vaccine research' *Alternative to Laboratory Animals* 2008; 36: 381–428.
107. Ledford H. 'HIV vaccine may raise risk' *Nature* November 15, 2007; 450 (7168): 325.
108. Feuer C, Bass E. 'AIDS vaccine update: Disappointment and questions after candidate fails in the STEP study' *The Body* July–December 2007. www.thebody.com, date accessed January 11, 2011.
109. Connor S, Green C. 'Is it time to give up the search for an AIDS vaccine?' *Independent* April 24, 2008.
110. Bailey J. 'Non-human primates in medical research and drug development: A critical review' *Biogenic Amines* 2005; 19: 235–255.
111. Schmidt C. 'Researchers exploring faster alternatives to 2-year test for carcongenicity' *Journal of the National Cancer Institute* 2006; 98: 228–230; Ward JM. 'The two-year rodent carcinogenesis bioassay—Will it survive?' *Journal of Toxicologic Pathology* 2007; 20: 13–19.
112. Gori GB. 'The costly illusion of regulating unknowable risks' *Regulatory Toxicology and Pharmacology* 2001; 34: 205–212; Ward, 2007; Schmidt, 2006; Ennever FK, Lave LB. 'Implications of the lack of accuracy of the lifetime rodent bioassay for predicting human carcinogenicity' *Regulatory Toxicology and Pharmacology* 2003; 38: 52–57.
113. Schmidt, 2006.
114. Ibid.
115. Ibid.
116. Ennever and Lave, 2003.
117. Knight A, Bailey J, Balcombe J. 'Animal carcinogenicity studies: 1. Poor human predictivity' *Alternative to Laboratory Animals* 2006; 34: 19–27.
118. Schmidt, 2006.
119. Ibid.
120. Heywood R. 'Target organ toxicity' *Toxicology Letters* 1981; 8: 349–358.
121. Allen, 2006.
122. Gori, 2001.
123. Fletcher AP. 'Drug safety tests and subsequent clinical experience' *Journal of the Royal Society of Medicine* 1978; 71: 693–696.
124. Allen, 2006.
125. Fletcher, 1978.

126. Neyt et al., 1998.
127. Brown LP. 'Do rats comply with EPA policy on cancer risk assessment for formaldehyde?' *Regulatory Toxicology and Pharmacology* 1989; 10: 196–200; Royal Society of Chemistry, Environment, Health and Safety Committee (EHSC). *Note on potency of chemical carcinogens* October 22, 2004. www.rsc. org, date accessed January 10, 2010.
128. Corry DB, Irvin CG. 'Promise and pitfalls in animal-based asthma research: Building a better mousetrap' *Immunology Research* 2006; 35: 279–294.
129. Contopoulis-Ioannadis DG, Ntzani EE, Ioannidis JPA. 'Translation of highly promising basic science research into clinical applications' *American Journal of Medicine* 2003; 114: 477–484.
130. Ioannidis, 2006.
131. Harris RBS. 'Appropriate animal models for clinical studies' *Annals of the New York Academy of Sciences* 1997; 819: 155–168; Dirnagl, 2006; Heywood, 1981; Heywood R. 'Animal models: Their use and limitations in long-term safety evaluation of fertility-regulating agents' *Human Reproduction* 1986; 1: 397–399; Tator CH. 'Review of treatment trials in human spinal cord injury: Issues, difficulties, and recommendations' *Neurosurgery* 2006; 59: 957–982; Heywood R. 'Target organ toxicity II' *Toxicology Letters* 1983; 18: 83–88; Sinha, 2006.
132. Corpet DE, Pierre F. 'How good are rodent models of carcinogenesis in predicting efficacy in humans? A systematic review and meta-analysis of colon chemoprevention in rats, mice and men' *European Journal of Cancer* 2005; 41: 1911–1922; Beyer C, Schett G, Distler O, Distler JHW. 'Animal models of systemic sclerosis' *Arthritis and Rheumatism* 2010; 62: 2831–2844; Lazzarini L, Overgaard KA, Conti E, Shirtliff ME. 'Experimental osteomyelitis: What have we learned from animal studies about the systemic treatment of osteomyelitis?' *Chemotherapy* 2000; 18: 451–460; Corry and Irvin, 2006; Krug N, Rabe KF. 'Animal models for human asthma: The perspective of a clinician' *Current Drug Targets* 2008; 9: 438–442; Gerlach M, Riederer P. 'Animal models of Parkinson's disease: An empirical comparison with the phenomenology of the disease in man' *Journal of Neural Transmission* 1996; 103: 987–1041; Codarri L, Fontana A, Becher B. 'Cytokine networks in multiple sclerosis: Lost in translation' *Current Opinions in Neurology* 2010; 23: 205–211; Jay et al., 2010; Tabakoff B, Hoffman PL. 'Animal models in alcohol research' *Alcohol Research and Health* 2000; 24: 77–84; Dyson A, Singer M. 'Animal models of sepsis: Why does preclinical efficacy fail to translate to the clinical setting?' *Critical Care Medicine* 2009; 37 (Suppl.): S30–S37; Marshall JC, Deitch E, Moldawer LL et al. 'Preclinical models of shock and sepsis: What can they tell us?' *Shock* 2005; 24 (Suppl. 1): 1–6; van der Staay FJ. 'Animal models of behavioral dysfunctions: Basic concepts and classifications, and an evaluation strategy' *Brain Research Reviews* 2006; 52: 131–159; New Scientist, 2004.
133. Knight A. 'Systematic reviews of animal experiments demonstrate poor contributions toward human healthcare' *Reviews on Recent Clinical Trials* 2008; 3: 89–96; Craig W. 'Relevance of animal models for clinical treatment' *European Journal of Clinical Microbiology and Infectious Diseases* 1993; 12 (Suppl. 1): S55–S57; Knight BA. 'Animal experiments scrutinised: Systematic review demonstrate poor human clinical and toxicological utility'

ALTEX 2007; 24: 320–325; Rohra DK, Jawaid A, Tauseef-ur-Rehman, Zaidi AH. 'Reliability of rodent animal models in biomedical research' *Journal of the College of Physicians and Surgeons Pakistan* 2005; 15: 809–812; Fletcher, 1978; Bailey, 2010; Bailey, 2008; Schein PS. 'Preclinical toxicology of anti-cancer agents' *Cancer Research* 1977; 37: 1934–1937; Lindl T, Voelkel M, Kolar R. 'Animal experiments in biomedical research. An evaluation of the clinical relevance of approved animal experimental projects' *ALTEX* 2005; 22: 143–151; Perel P, Roberts I, Sena E et al. 'Comparison of treatment effects between animal experiments and clinical trials: Systemic review' *British Medical Journal* Published online December 15, 2006. doi: 10.1136/bmj.39048.407928.BE.

134. Bracken MB. 'Why are so many epidemiology associations inflated or wrong? Does poorly conducted animal research suggest implausible hypotheses?' *Annals of Epidemiology* 2009; 19: 220–224.

135. Francione GL. 'The use of nonhuman animals in biomedical research: Necessity and justification' *Journal of Law, Medicine and Ethics* 2007; Summer: 241–248.

136. Lemon R, Dunnett SB. 'Surveying the literature from animal experiments. Critical reviews may be helpful—Not systematic ones' *British Medical Journal* 2005; 330: 977–978.

137. Roberts et al., 2002.

138. Sams-Dodd, 2006.

139. Wall and Shani, 2008.

140. The Plain Dealer. 'Drug-testing labyrinth' *The Plain Dealer* January 10, 1995.

141. Hartung T. 'Per aspirin ad astra ...' *ATLA* 2009; 37 (Suppl. 2): 45–57.

142. Greek R, Greek J. 'Animal research and human disease' *Journal of the American Medical Association* 2000; 283: 743–744.

143. Coleman, 2010.

144. Allen, 2006.

145. Monastersky R. 'Protesters fail to slow animal research' *Chronicle of Higher Education* 2008; 54: 1.

146. American Association for the Advancement of Science. 'NIH budget flat in 2009 proposal' February 19, 2008. www.aaas.org/spp/rd, date accessed December 10, 2010.

147. US Department of Agriculture, Animal and Plant Health Inspection Service. 'Annual report animal usage by fiscal year.' www.aphis.usda.gov, date accessed July 6, 2010.

148. Sullivan M. 'The Animal Welfare Act—What's that?' *NYSBA Journal* July/August, 2007: 17–23; Mukerjee M. 'Speaking for the animals: A veterinarian analyzes the turf battles that have transformed the animal laboratory' *Scientific American* August 2004.

149. Canadian Council on Animal Care. '2009 CCAC Survey of Animal Use' Ottawa, Ontario 2010.

150. Anonymous. 'Humane league' *Economist* 2007; 384: 70–71.

151. Anonymous. *Guide for the Care and Use of Laboratory Animals*, 8th edn. (Washington, DC: The National Academies Press) 2011. Emphasis added.

152. Walker RL. 'Human and animal subjects of research: The moral significance of respect versus welfare' *Theoretical Medicine and Bioethics* 2006; 27: 305–331.

153. Orlans FB. 'Ethical decision making about animal experiments' *Ethics & Behavior* 1997; 7: 163–171; Rollin BE. 'The regulation of animal research and the emergence of animal ethics: A conceptual history' *Theoretical Medicine and Bioethics* 2006; 27: 285–304; Walker, 2006; Kolar R. 'Animal experimentation' *Science and Engineering Ethics* 2006; 12: 111–122.
154. Sullivan, 2007.
155. Ibid.
156. Flecknell P. 'Partnerships for progress' *Veterinary Anaesthesia and Analgesia* 2005; 32: 239–240.
157. Prescott MJ, Morton DB, Anderson D et al. 'Refining dog husbandry and care—Eighth report of BVAAWF/FRAME/RSPCA/UFAW—Joint Working Group on Refinement' *Laboratory Animals* 2004: 38 (Suppl. 1): S1–S94; Würbel H, Stauffacher M, von Holst D. 'Stereotypies in laboratory mice— Quantitative and qualitative description of the ontogeny of "wire-gnawing" and "jumping" in Zur: ICR and Zur: ICR nu' *Ethology* 1996; 102: 371–385; Lutz C. 'Stereotypic and self-injurious behaviour in rhesus macaques: A survey and retrospective analysis of environment and early experience' *American Journal of Primatology* 2003; 60: 1–15; Novak MA. 'Self-injurious behaviour in rhesus monkeys: New insights into its etiology, physiology, and treatment' *American Journal of Primatology* 2003; 59: 3–19.
158. Mason GJ, Latham NR. 'Can't stop, won't stop: Is stereotypy a reliable animal welfare indicator?' *Animal Welfare* 2004; 13 (Suppl. 1): 57–69 (13).
159. Vertein R, Reinhardt V. 'Training female rhesus monkeys to cooperate during in-homecage venipuncture' *Laboratory Primate Newsletter* 1989; 28: 1–3.
160. Suckow MA, Weisbroth SH, Franklin CL (eds) *The Laboratory Rat*, 2nd edn. (Burlington, MA: Elselvier Academic Press) 2006, p. 323.
161. Balcombe et al., 2004.
162. Suckow et al., 2006., p. 323.
163. Balcombe et al., 2004.
164. Flow BL, Jaques JT. 'Effect of room arrangement and blood sample collection sequence on serum thyroid hormone and cortisol concentrations in cynomolgus macaques (*Macaca fascicularis*)' *Contemporary Topics in Laboratory Animal Science* 1997; 36: 65–68.
165. Balcombe et al., 2004.
166. Nakayama K, Goto S, Kuroaka K, Nakamura K. 'Decrease in nasal temperature of rhesus monkeys (*Macaca mulatta*) in negative emotional state' *Physiology & Behavior* 2005; 84: 783–790.
167. Bekoff M. 'Animal welfare' *New Scientist* 2009; 2690: 17; Sullivan, 2007.
168. United States Department of Agriculture, Office of the Inspector General. 'Audit Report'. *APHIS Animal Care Program Inspection and Enforcement Activities* September 2005.
169. Clemedson CE, McFarlane-Abdulla M, Andersson FA et al. 'MEIC evaluation of acute systemic toxicity. Part II. *In vitro* results from 68 toxicity assays used to test the first 30 reference chemicals and a comparative cytotoxicity analysis' *ATLA* 1996; 24: 273–311; Clemedson CE, Barile FA, Chesne C et al. 'MEIC evaluation of acute systemic toxicity. Part VII. Prediction of human toxicity by results from testing of the first 30 reference chemicals with 27 further *in vitro* assay' *ATLA* 2000; 28: 161–200.

170. MatTek Corporation. Reports presented at the annual Society of Toxicology meeting in Seattle, held March 16–20, 2008. www.mattek.com, date accessed July 20, 2008.

171. Voskoglou-Nomikos T, Pater JL, Seymour L. 'Clinical predictive value of the *in vitro* cell line, human xenograft, and mouse allograft preclinical cancer models' *Clinical Cancer Research* 2003; 9: 4227–4239.

172. Mayor S. 'Researchers refine in vitro test that will reduce risk of "first in humans" drug trial' *British Medical Journal* 2009; 358: 11.

173. Coleman, 2010.

174. Arnaud CH. 'Systems biology's clinical future. Although it now remains a research tool, systems biology is moving toward clinical applications, including personalized medicine' *Chemical and Engineering News* July 31, 2006; 84: 17–26.

175. Coghlan A. 'Pioneers cut out animal experiments' *New Scientist* August 31, 1996; 2045.

176. Ibid.

177. University of Miami, Miller School of Medicine. *The Miami Project to Cure Paralysis.* www.themiamiproject.org, date accessed January 13, 2011.

178. Leslie, 2010.

179. Howard Hughes Medical Institute. 'Scientists and research'. www.hhmi.org, date accessed January 16, 2011.

180. BBC News. 'Human metabolism recreated in lab' January 30, 2007. news.bbc.co.uk, date accessed December 29, 2010.

181. National Research Council of the National Academies. *Toxicity Testing in the 21st Century: A Vision and a Strategy* (Washington, DC: The National Academies Press) 2007.

182. DeGrazia D, Ashcroft RE, Dawson A, Draper H, McMillan J. (eds) *On the Ethics of Animal Research in Principles of Health Care Ethics*, 2nd edn. (West Sussex, England: Wiley Publications) 2007, p. 692.

183. Goldberg AM, Hartung T. 'Protecting more than animals: Reducing animal suffering often has the unexpected benefit of yielding more rigorous safety tests' *Scientific American* 2006; 294: 84–91; Becker RA, Borgert CJ, Webb S et al. 'Report of an ISRTP Workshop: Progress and barriers to incorporating alternative toxicological methods in the U.S.' *Regulatory Toxicology and Pharmacology* 2006; 46: 18–22.

184. Becker et al., 2006.

185. Hartung T. 'Food for thought … on animal tests' *ALTEX* 2008; 25: 3–9.

186. National Institutes of Health. 'Computer Retrieval on Information on Scientific Projects'. http://crisp.cit.nih.gov, date accessed January 11, 2008; The Sunshine Project. 'Computer Retrieval on Information on Scientific Projects—Extended Report'. www.sunshine-project.org/crisper, date accessed January 11, 2008.

187. Gottschall JG, Nichols R. 'Head pitch affects muscle activity in the decerebrate cat hindlimb during walking' *Experimental Brain Research* 2007; 182: 131–135.

188. Ballard CL, Wood RI. 'Partner preference in male hamsters: Steroids, sexual experience and chemosensory cues' *Physiology and Behavior* 2007; 91: 1–8.

189. Brenowitz EA, Lent K, Rubel EW. 'Auditory feedback and song production do not regulate seasonal growth of song control circuits in adult white-crowned sparrows' *Journal of Neuroscience* 2007; 27: 6810–6814.

190. Alekseyenko OV, Waters P, Zhou H, Baum MJ. 'Bilateral damage to the sexually dimorphic medial preoptic area/anterior hypothalamus of male ferrets causes a female-typical preference for and a hypothalamic Fos response to male body odors' *Physiology and Behavior* 2007; 90: 438–449.

191. de Oca BM, Minor TR, Fanselow MS. 'Brief flight to a familiar enclosure in response to a conditional stimulus in rats' *Journal of General Psychology* 2007; 134: 153–172.

192. Winger G, Galuska CM, Hursh SR. 'Modification of ethanol's reinforcing effectiveness in rhesus monkeys by cocaine, flunitrazepam, or gamma-hydroxybutyrate' *Psychopharmacology* 2007; 193: 587–598.

193. Rollin BE. 'Animal research: A moral science' *EMBO Reports* 2007; 8: 521–525.

194. Greek R, Greek J. 'Is the use of sentient animals in basic research justifiable?' *Philosophy, Ethics, and Humanities in Medicine* 2010; 5: 14.

7   The New Public Health

1. Schweitzer A. Nobel lecture 'The problem of peace' November 4, 1954. www.nobelprize.org, date accessed January 13, 2010.

2. Layton L. 'Wal-Mart turns to "retail regulation" to ban flame retardant' *Washington Post* February 25, 2011.

3. Appiah KA. 'What will future generations condemn us for?' *Washington Post* September 26, 2010.

Index

accelerating virus mutation, vaccinations and the risk of, 114–15

adolescent sadism, Johnson and Becker's case series, 31

Africa
unusual die-off in pigs prior to deadly Ebola outbreaks, 111
zoonotic pathogens emerging from, 68

African dwarf frogs, 74, 83

African trypanosomiasis, 81

Agnes Grey (Brontë), 27

agricultural subsidies, health impact of policy changes, 130

AIDS, 61, 64
see also HIV/AIDS

alcohol addiction, 153

algae, 118

altruism, of vampire bats, 10

Alzheimer's disease, animal models of, 141–2, 147

Amazing Animal Productions, 56

American Dietetic Association, 128

American Humane Association, 39

American Pet Product Manufacturers Association, 56

amphibians
global threat to, 78
Salmonella in, 74

amyotrophic lateral sclerosis (Lou Gehrig's disease), 146

animal abuse
children and, 39, 42
and criminal behaviour, 30–3
defining, 45
as indicator of mental illness, 46
NSPCC findings, 46–7
peer bullying study, 42
public health perspective, 47–8
reasons for, 34–5, 44

as a 'red flag' for other forms of violence, 47
scarcity of official data on, 45

animal agriculture
energy input *vs.* output, 121–2
estimated global burden of disease attributable to, 128–9
global land use, 121
greenhouse gases produced by, 120–1
industrialisation, 87
as major driver of biodiversity loss, 121
overview, 87–8
pollution caused by, 118–19
strategies to reduce the carbon footprint of, 124
see also factory farming; intensive farming practices

animal behaviour, revolution in the field of, 8

animal experience, experimental evidence, 12–14

animal experimentation
amount spent on, 156
aspirin, 155
barriers to development of alternatives, 162–3
concerns about the reliability of, 136
costs to animals, 156–8
developing alternatives to, 158–63
and the discovery of a second thymus gland in mice, 144
and drug development failure, 133–4, 139–41, 146, 148–9, 155, 166
dubious experiments, 163–5
and environmental influences, 137–9
financial investments, 162
genetically engineered model, 146–7

CPSIA information can be obtained at www.ICGtesting.com
Printed in the USA
LVOW07*2131310713

345716LV00008B/171/P

9 780230 249738

Libraries @ Becker College